Control of nonlinear differential algebraic equation systems

Submission of proposals for consideration
Suggestions for publication, in the form of outlines and representative samples, are invited by the Editorial Board for assessment. Intending authors should approach one of the main editors or another member of the Editorial Board, citing the relevant AMS subject classifications. Alternatively, outlines may be sent directly to the publisher's offices. Refereeing is by members of the board and other mathematical authorities in the topic concerned, throughout the world.

Preparation of accepted manuscripts
On acceptance of a proposal, the publisher will supply full instructions for the preparation of manuscripts in a form suitable for direct photo-lithographic reproduction. Specially printed grid sheets can be provided. Word processor output, subject to the publisher's approval, is also acceptable.

Illustrations should be prepared by the authors, ready for direct reproduction without further improvement. The use of hand-drawn symbols should be avoided wherever possible, in order to obtain maximum clarity of the text.

The publisher will be pleased to give guidance necessary during the preparation of a typescript and will be happy to answer any queries.

Important note
In order to avoid later retyping, intending authors are strongly urged not to begin final preparation of a typescript before receiving the publisher's guidelines. In this way we hope to preserve the uniform appearance of the series.

CRC Press UK
Chapman & Hall/CRC Statistics and Mathematics
Pocock House
235 Southwark Bridge Road
London SE1 6LY
Tel: 0171 407 7335

Aditya Kumar
University of Minnesota

Prodromos Daoutidis
University of Minnesota

Control of nonlinear differential algebraic equation systems

with applications to chemical processes

CHAPMAN & HALL/CRC

Boca Raton London New York Washington, D.C.

Library of Congress Cataloging-in-Publication Data

Kumar, Aditya.
    Control of nonlinear differential algebraic equation systems /
    Aditya Kumar and Prodromos Daoutidis.
        p. cm. -- (Chapman & Hall/CRC research notes in mathematics)
    Includes bibliographical references and index.
    ISBN 0-8493-0609-4 (alk. paper)
      1. Control theory. 2. Differential algebraic equations.
    3. Nonlinear theories. 1. Kumar, Aditya. II. Title. III. Series:
    Chapman & Hall/CRC research notes in mathematics series.
    QA402.3.D36 1999
    629.8'312--dc21   98-32277
                                                CIP

The text in Sections 4.3–4.5 and 7.4.1 are reprinted from the *AIChE Journal*, August 1996 with permission of the American Institute of Chemical Engineers. Copyright © 1996 AIChE. All rights reserved.

Portions of the text in Section 6.3 are reprinted from *Chemical Engineering Science*, Volume 53, A. Kumar and P. Daoutidis, Singular Perturbation Modeling to Nonlinear Processes with Non-Explicit Time-Scale Multiplicity, Pages No. 1491–1504, Copyright (1998), with permission from Elsevier Science.

Contents

Preface

Many engineering applications are modeled by systems of coupled differential and algebraic equations (DAEs) that cannot be directly reduced to ordinary differential equations (ODEs). Such systems (also referred to as singular, implicit etc.) have been studied extensively from the point of view of numerical simulation. The feedback control of linear DAE systems has also been studied extensively. On the other hand, the feedback control of nonlinear DAE systems is a relatively new subject, that has been developing steadily over the last few years, inspired, in part, by research advances in the feedback control of nonlinear ODE systems.

This research note presents, in a unified framework, recent results on the stabilization, output tracking and disturbance elimination for a large class of nonlinear DAE systems. The intent is not to include all of the developments on this subject, but, through a focused exposition, to introduce the reader to the inherent characteristics of such systems, and the tools and methods that we can employ to address their control. A second objective is to document the occurrence and significance of DAE systems in the modeling and control of chemical processes, a large class of applications that have received limited attention in the existing literature on singular systems (in contrast to electrical circuits and mechanical systems).

The book is intended primarily as a reference for engineers and applied mathematicians interested in the analysis and control of dynamical systems and the mathematical modeling of engineering applications. It is written at a basic mathematical level, assuming some familiarity with concepts in analysis and control of nonlinear ODE systems.

More specifically, the book focuses on continuous-time systems of differential and algebraic equations in semiexplicit form (the most common form for the applications in mind). Chapter 1 provides a discussion of results on the analysis and numerical simulation of DAE systems and the control of linear DAE systems. Emphasis is placed on illustrating the fundamental differences between DAE and ODE systems, their implications in simulation and control, and the available approaches to address these issues. The concepts of index (a measure of the "singularity" of the DAE system) and regularity (the existence of a control-invariant state space) that play an important role in the subsequent chapters are also introduced and discussed. Chapter 2 describes examples of generic classes of chemical processes with fast rates of mass transfer, heat transfer, reaction or gaseous flow, that are naturally modeled by high-index DAE systems under the (quasi-steady-state) assumptions of phase, thermal or reaction equilibrium, and negligible pressure drop. The occurrence of high-index DAE models in networks (interconnections) of chemical processes is also documented.

Chapter 3 addresses the derivation of state-space realizations of regular high-index DAE systems, and the synthesis of state feedback controllers for stabilization and tracking on the basis of the state-space realizations. Chapter 4 deals with nonregular DAE systems, for which a state-space realization does not exist independently of the controller design. The controller synthesis for such systems is addressed through a regularizing feedback modification and the derivation of state-space realizations of the feedback regularized system. Chapter 5 focuses on high-index DAE systems with disturbance inputs, and addresses the feedforward/state feedback control of systems that possess a well-defined state-space realization either independently of the disturbances, or if they are measured, after a feedforward/feedback regularization. Chapter 6 addresses the connections between the high-index DAE systems considered in this work and a class of singularly perturbed systems in nonstandard form. The results provide a rigorous justification for the quasi-steady-state approximations and the resulting high-index DAE models of the chemical processes presented in Chapter 2. Finally, Chapter 7 includes several simulation case studies on representative chemical processes, that illustrate the application of the control methods and the advantages of using the (quasi-steady-state) high-index DAE models as the basis for the controller design.

We are grateful to our colleagues in the Department of Chemical Engineering and Materials Science at the University of Minnesota, for having created a stimulating atmosphere of academic excellence, within which the research that led to this book was performed over the last five years. We are particularly indebted to Rutherford Aris for his encouragement and advice in the various phases of the preparation of this book. We are also indebted to Steve Campbell for his constructive comments and suggestions. Finally, we would like to thank the National Science Foundation which provided funding (through grant no. CTS-9320402) for most of the research that formed the basis for this book.

I would like to express my gratitude to Prodromos for his constant encouragement, especially during the last year. I am forever grateful to my family for their affection and unquestioning support. The presence of my wife Riti beside me, made the completion of this book all the more gratifying. I dedicate this book to fond memories of my mother, who always found joy in my endeavors.

A.K.

I owe a special note of gratitude to my wife Aphrodite, for her unselfish and generous devotion to our common life during the last five years. I dedicate this book to her and our son Stylianos.

P.D.

Minneapolis
September 11, 1998

1. DAEs: Background and Concepts

1.1 Introduction

Differential algebraic equation (DAE) systems (also referred to as singular, implicit, descriptor, semistate, and generalized systems) arise naturally as dynamic models of chemical [11,90], electrical [12,107], and mechanical engineering [60,101,106] applications. It is now well-established that DAE systems may have fundamentally different characteristics from ordinary differential equation (ODE) systems [115]. For instance, unlike ODEs, arbitrary initial conditions or nonsmooth inputs in DAEs may lead to impulsive solutions, which is a common cause for failure of standard ODE simulation methods. A concept that is commonly used to provide a measure of the differences between DAE and ODE systems is that of the (differential) index. Loosely speaking, the index of a DAE system is the minimum number of differentiations required to obtain an equivalent ODE system. DAE systems with index exceeding one are referred to as high-index systems. Such high-index DAEs have been a subject of extensive research, with particular emphasis on their numerical analysis and simulation (see the books [9,57] and the references therein). A variety of numerical simulation methods have been developed for specific classes of high-index DAEs, ranging from methods using a combination of index-reduction techniques with efficient ODE integration methods for the resulting index-one or index-zero DAE (e.g., [2,8,30,48,50]), to those employing nonlinear constrained optimization techniques [67,124]. The development of a numerical simulation method for general high-index DAEs is still an active area of research (see e.g., [19,20]).

Early research on the control of DAE systems focused on linear systems (see e.g., [14,37,95] for a survey of results). For such systems, fundamental properties like solvability and stability, and solution characteristics (e.g., [12,32]), and control-relevant concepts like controllability and observability (e.g., [14,34,103,138]) system equivalence (e.g., [58,117,128,134,139]) and minimal realizations (e.g., [26,56,81,104]) have been studied. These results have led to a variety of controller design methods, which can be broadly categorized into two classes: (i) those within the framework of classical smooth solutions corresponding to "consistent" initial conditions (e.g., [14,76, 89]), and (ii) those corresponding to "arbitrary" initial conditions (e.g., [6,10,31,33,34, 68,97,134]) where the removal of impulsive behavior through a smooth state/output feedback is an additional problem to be addressed in the controller design.

1

The progress in the control of linear DAE systems and the advances in the control of nonlinear ODE systems (see e.g., [64, 70, 80, 108]) have spurred a steadily growing research activity on the control of nonlinear DAE systems over the last decade. In this regard, system-theoretic properties like existence and uniqueness of solutions (e.g. [18,43,96,121,123,125]), and stability analysis using Lyapunov techniques [4,5,43] have been studied for different classes of nonlinear DAE systems. A variety of controller design results have also been derived (e.g., [25,75,77,79,87,88,91,102,136,137]). These results lie within the classical framework of smooth solutions, and in general, rely on the derivation of an ODE representation of the original or a feedback modified DAE system. Finally, the design of optimal controllers through nonlinear programming has also been addressed for certain classes of nonlinear DAE systems [35,36].

In this chapter, we first review, briefly, the available results for linear DAE systems. Thereafter, we introduce the general class of nonlinear DAE systems that we focus on throughout this research note and review concepts like solvability and index for such systems. We discuss the key differences between ODE and high-index DAE systems that lead to well-known problems in the numerical simulation of DAEs. Finally, we present a notion of regularity [88], which is used in the later chapters to distinguish between classes of nonlinear high-index DAE systems that are fundamentally different in the context of feedback control.

1.2   Linear Systems

Consider a linear system with the following description:

$$E\dot{x} = Ax + Bu(t) \tag{1.1}$$

where $x \in \mathbb{R}^n$ is the vector of state variables, $u(t) \in \mathbb{R}^m$ is the vector of input variables, and $E, A \in \mathbb{R}^{n \times n}$ and $B \in \mathbb{R}^{n \times m}$ are constant matrices. Clearly, if the matrix E is nonsingular, then the above system is equivalent to a standard ODE system:

$$\dot{x} = \bar{A}x + \bar{B}u(t) \tag{1.2}$$

where $\bar{A} = E^{-1}A$ and $\bar{B} = E^{-1}B$. However, if E is singular then the system of Eq.1.1 is different from an ODE system.

Systems of the form of Eq.1.1 with a singular matrix E are often referred to as singular or generalized state-space systems (see, e.g., [134]). The solution characteristics of the singular system in Eq.1.1 are determined by the corresponding matrix pencil $\mathcal{P} = (sE - A)$ [47], where $s \in \mathbb{C}$. More specifically, the system is solvable if and only if the pencil $\mathcal{P}$ is regular, i.e., the polynomial $\det(\mathcal{P})$ is not identically zero; a practical procedure for verifying the regularity of the matrix pencil is provided by Luenberger's shuffle algorithm [98].

The solution characteristics of a solvable DAE system of Eq.1.1 become apparent in a canonical form representation. More specifically, if $\text{rank}(E) = r < n$ and $\det(\mathcal{P})$ is a (nonzero) polynomial of degree s $(0 \le s \le r)$, then there exist nonsingular matrices $P, Q \in \mathbb{R}^{n \times n}$ such that premultiplying Eq.1.1 with P and employing a coordinate change:

$$\bar{x} = \begin{bmatrix} \bar{x}_1 \\ \bar{x}_2 \end{bmatrix} = Qx$$

yields the following standard canonical form representation [47]:

$$\begin{aligned} \dot{\bar{x}}_1 &= A_1 \bar{x}_1 + B_1 u(t) \\ N\dot{\bar{x}}_2 &= \bar{x}_2 + B_2 u(t) \end{aligned} \qquad (1.3)$$

where $\bar{x}_1 \in \mathbb{R}^s$ and $\bar{x}_2 \in \mathbb{R}^{n-s}$. In the above description, A_1, B_1, and B_2 are constant matrices of appropriate dimensions, and N is an $(n-s) \times (n-s)$ matrix of nilpotency ν, i.e., all eigenvalues of N are zero and $N^\nu = 0$ while $N^i \neq 0$ for $i \le \nu - 1$. In the specific case when $s = r$, the matrix N is identically zero with an index of nilpotency $\nu = 1$, and the $\bar{x}_2$ subsystem in Eq.1.3 is a purely algebraic system. On the other hand, if $s < r$ and N is in the Jordan canonical form, then the index $\nu > 1$ is the size of the largest Jordan block.

In the system of Eq.1.3, the ODE subsystem in $\bar{x}_1$ is decoupled from the $\bar{x}_2$ subsystem and has the standard solution:

$$\bar{x}_1(t) = e^{A_1 t} \bar{x}_1(0) + \int_0^t e^{A_1(t-\tau)} B_1 u(\tau) d\tau, \quad t \ge 0 \qquad (1.4)$$

for any initial condition $\bar{x}_1(0)$ and continuous inputs $u(t)$. While the solution for $\bar{x}_1(t)$ depends on the choice of the initial condition $\bar{x}_1(0)$, the solution for $\bar{x}_2(t)$, including the initial condition $\bar{x}_2(0)$, is uniquely determined by the forcing inputs $u(t)$:

$$\bar{x}_2(t) = -\sum_{i=0}^{\nu-1} N^i B_2 u^{(i)}(t), \quad t \ge 0 \qquad (1.5)$$

where $u^{(i)}(t)$ denotes the ith derivative of the input. Thus, unlike ODE systems, DAE systems of Eq.1.1 do not have smooth solutions for arbitrary initial conditions. Only initial conditions $x(0)$ that are consistent, i.e., $x(0)$ for which $\bar{x}_2(0)$ satisfy Eq.1.5 at $t = 0$, yield smooth solutions. Furthermore, unlike ODE systems, if the index ν for the DAE system of Eq.1.1 exceeds one, then the solution of the DAE system may depend on the derivatives of the forcing inputs $u(t)$, which must be accordingly smooth.

Remark 1.1: The solvability of linear time-invariant systems of Eq.1.1 is characterized completely in terms of the associated matrix pencil $\mathcal{P}$, i.e., solvability is equivalent to the regularity of the matrix pencil. However, for linear time-varying systems with the description:

$$E(t)\dot{x}(t) = A(t)x(t) + B(t)u(t), \qquad (1.6)$$

3

solvability of the system and regularity of the matrix pencil $\mathcal{P} = (sE(t) - A(t))$ are totally independent concepts [9]. A characterization of solvability for linear time-varying systems of Eq.1.6, and a general canonical form representation for solvable systems were obtained in [16].

The lack of classical smooth solutions for arbitrary initial conditions in a DAE system of Eq.1.1 prompted research on studying the existence of solutions in a more general distributional sense. This was motivated by the consideration that the singularity of the matrix E may arise from a sudden change in the dynamic system (e.g., a switch closing/opening in an electrical circuit) or from approximating some small parameters with zero, and thus, the state x of the system at time $t = 0_-$ need not be constrained. For such an arbitrary initial condition $x(0_-) \in \mathbb{R}^n$, the system of Eq.1.1 has a distributional solution under the same condition of regularity of the matrix pencil $\mathcal{P}$. More specifically, for the system in Eq.1.3 with arbitrary initial condition $\bar{x}_2(0_-)$, the $\bar{x}_2$ subsystem has the following distributional solution [32, 133]:

$$\bar{x}_2(t) = -\sum_{i=1}^{\nu-1} \delta^{(i-1)} N^i \bar{x}_2(0_-) - \sum_{i=0}^{\nu-1} N^i B_2 u^{(i)}(t) , \quad t \geq 0 \tag{1.7}$$

where δ denotes the unit impulse (Dirac delta) function and $\delta^{(i)}$ denotes the ith distributional derivative. Starting from an arbitrary initial condition $\bar{x}_2(0_-)$, the above solution for $\bar{x}_2(t)$ exhibits an impulsive behavior at $t = 0$, and thereafter, it is the same as the classical solution in Eq.1.5 corresponding to consistent initial conditions.

The impulsive behavior in the solution for $\bar{x}_2$ essentially corresponds to the presence of poles at infinity in the system of Eq.1.1. More specifically, in a generalized state-space framework [134], the system of Eq.1.1, or equivalently Eq.1.3, has r dynamic modes and $n - r$ nondynamic or algebraic modes, where the dynamic modes include s finite modes corresponding to the finite poles and $r - s$ impulsive modes corresponding to poles at infinity. Clearly, the system of Eq.1.1 has impulsive modes if and only if $\nu > 1$, or equivalently $s < r$, i.e., the degree of the polynomial $\det(sE - A)$ is strictly less than the rank of the matrix E.

The presence of impulsive modes in the system of Eq.1.1 is not desirable, and thus, research on the control of singular systems with arbitrary initial conditions focused on eliminating the impulsive behavior through smooth state feedback. This is possible for systems that are controllable at infinity [6, 34], or equivalently systems for which the pencil $(sE - A - BK)$ has index $\nu = 1$ for some feedback $u = Kx + v$ (i.e., the polynomial $\det(sE - A - BK)$ has the same degree as the rank of the matrix E). For such systems, the problem of feedback pole placement [31] and optimal control [6, 33, 68] through state feedback has been addressed. Under analogous conditions of observability of the finite and impulsive modes (termed as impulse observability or observability at infinity) [34], the design of state observers for the system of Eq.1.1 with outputs $y = Cx + Du$ has also been addressed [1, 63, 112, 129]. Furthermore, the removal of impulsive behavior [10, 97] and pole placement [51, 52] through output feedback

4

have been addressed under similar conditions. The analysis and control results for linear time-invariant continuous systems of Eq.1.1 have also been generalized to time-varying [14] and discrete-time [37] systems.

A key requirement in the above-mentioned results is that the DAE system of Eq.1.1 must be controllable at infinity. However, there is a broad class of DAE systems that are not controllable at infinity, for which it is not possible to eliminate the impulsive behavior arising from arbitrary initial conditions. Thus, another research direction has focused on addressing the control of DAE systems within the conventional perspective of smooth solutions corresponding to consistent initial conditions [14, 76, 89].

1.3 Nonlinear Systems

1.3.1 Descriptions

Research on the analysis and numerical simulation of nonlinear DAEs has focused on systems with the following general *fully-implicit* description:

$$F(\dot{x}, x, u(t)) = 0 \tag{1.8}$$

where $x \in \mathcal{X} \subset \mathbb{R}^n$ is the vector of state variables ($\mathcal{X}$ is an open connected set), $u(t) \in \mathbb{R}^m$ is the vector of input variables and $F : \mathbb{R}^n \times \mathcal{X} \times \mathbb{R}^m \to \mathbb{R}^n$ is a smooth function. In this chapter, we discuss the characteristics of DAE systems with specified time-varying inputs $u(t)$. In the subsequent chapters where we address the controller design, u denotes the vector of *manipulated* (or control) inputs that vary according to a feedback control law. Clearly, the system in Eq.1.8 is an implicit ODE system if the Jacobian $\partial F / \partial \dot{x}$ is nonsingular. On the other hand, if $\partial F / \partial \dot{x}$ is singular, then the system exhibits fundamentally different characteristics from ODE systems.

There has been a substantial amount of research on addressing basic system-theoretic issues like existence and uniqueness of solutions [18, 120, 121, 123], stability [4, 5, 99], etc., for fully-implicit DAE systems of Eq.1.8. However, the generality of the form of the system in Eq.1.8 does not allow the development of explicit controller synthesis results. Furthermore, the majority of chemical process applications (and other engineering applications as well) are modeled by DAE systems in the so-called *semiexplicit* form where there is a distinct separation of the differential and algebraic equations. In particular for chemical processes, the standard dynamic balances of mass and energy yield explicit differential equations, while thermodynamic relations, empirical correlations, quasi-steady-state relations, etc., comprise the algebraic equations.

Motivated by the above considerations, we will focus mainly on nonlinear DAE systems with the following semiexplicit description:

$$\begin{aligned} \dot{x} &= f(x) + b(x)z + g(x)u(t) \\ 0 &= k(x) + l(x)z + c(x)u(t) \end{aligned} \tag{1.9}$$

5

where $x \in \mathcal{X} \subset \mathbb{R}^n$ is the vector of differential variables for which we have the explicit differential equations, $z \in \mathcal{Z} \subset \mathbb{R}^p$ is the vector of algebraic variables that vary according to the algebraic equations, $\mathcal{X}$ and $\mathcal{Z}$ are open connected sets, $u(t) \in \mathbb{R}^m$ is the vector of input variables, $f(x)$ and $k(x)$ are smooth vector fields of dimensions n and p, respectively, and $b(x), g(x), l(x)$, and $c(x)$ are smooth matrices of appropriate dimensions. Note that in the above description, the inputs u and the algebraic variables z appear in the system equations in a linear (affine) fashion, which is typical of most practical applications. This linearity with respect to z and u, and the semiexplicit structure, will facilitate the analysis and the derivation of explicit controller synthesis results in the later chapters. Systems that are nonlinear in u and/or z can also be easily recast in the above form through standard dynamic extension techniques (see e.g., [108], page 190). For instance, consider the following DAE system that is nonlinear in z:

$$\dot{x} = f(x, z) + g(x, z)u(t)$$
$$0 = k(x, z) + c(x, z)u(t) \tag{1.10}$$

Defining the extended vector of differential variables $\bar{x} = [x^T \ z^T]^T$ and the new vector of algebraic variables $\bar{z} = \dot{z}$, the following extended DAE system is obtained:

$$\begin{bmatrix} \dot{x} \\ \dot{z} \end{bmatrix} = \begin{bmatrix} f(\bar{x}) \\ 0 \end{bmatrix} + \begin{bmatrix} 0 \\ I_p \end{bmatrix} \bar{z} + \begin{bmatrix} g(\bar{x}) \\ 0 \end{bmatrix} u(t)$$
$$0 = k(\bar{x}) + c(\bar{x})u(t) \tag{1.11}$$

which is clearly in the form of Eq.1.9 with $\bar{z}$ appearing in a linear fashion. Similarly, fully-implicit DAE systems of Eq.1.8 can also be transformed into a semiexplicit form by defining the algebraic variables $z = \dot{x}$ [48].

1.3.2 Solvability and Index

Initial research on solvability of nonlinear DAE systems focused on rather specialized classes of systems. More specifically, in [125], the solvability of a class of systems with the description:

$$F(y, t) = 0, \qquad A(y, t)\frac{dy}{dt} = G(y, t) \tag{1.12}$$

was studied using the theory of differential equations on manifolds. In Eq.1.12, F and G are vector-valued mappings of dimensions n and m $(m < n)$, respectively, and A is a matrix operator of dimension $m \times n$. Under sufficient smoothness of F, A, and G, and the condition that the matrix:

$$\begin{bmatrix} D_y F(y, t) \\ A(y, t) \end{bmatrix} \tag{1.13}$$

6

is nonsingular on the manifold where $F(y,t) = 0$ (in Eq.1.13, $D_y F(y,t)$ denotes the Jacobian $(\partial F(y,t)/\partial y)$, the system of Eq.1.12 corresponds to a locally unique vector field on this manifold. On the other hand, if the matrix in Eq.1.13 is singular, then the system in Eq.1.12 permits solutions only on a lower dimensional manifold. This latter class of systems was termed as algebraically incomplete; they are essentially high-index systems (a precise definition of index is given later in this section). The work of [125] was extended in [126] to a more general class of first- and second-order DAE systems with an index not exceeding two and three, respectively.

This differential-geometric approach of viewing DAE systems as appropriate vector fields on manifolds was used in [122] to address the solvability of DAE systems with the description:

$$A(z)\frac{dz}{dt} = f(z) \tag{1.14}$$

where the matrix $A(z)$ is singular and then generalized in [123] to more general systems of the form:

$$F(t, x, \dot{x}) = 0 \tag{1.15}$$

The approach for the specification of sufficient conditions for solvability entailed a recursive identification of a family of constraint manifolds M_i, $i = 0, \ldots s$, where $M_{i+1} \subset M_i$ and s, called the degree of the DAE system, is the largest integer such that $M_{s-1} \neq M_s$. The DAE system was termed as *regular* if, under appropriate conditions, it corresponded to a unique vector field on the constrained manifold, and thus, permitted a locally unique solution on this manifold. An analogous approach was followed in [120, 121] to derive sufficient conditions under which a general DAE system of the form in Eq.1.15 is equivalent to an ODE system on an appropriate constrained manifold. More specifically, for an autonomous (a nonautonomous system of Eq.1.15 is converted to an autonomous one using the standard trick of including t in an extended state vector and adding the differential equation $\dot{t} = 1$) DAE system:

$$F(x, \dot{x}) = 0 \tag{1.16}$$

a sequence of mappings $F^j(x, p)$, $j = 0, \ldots k + 1$ ($p \in \mathbb{R}^n$ is used to denote $\dot{x}$) are defined recursively through appropriate differentiations and orthogonal projections, until the Jacobian $F_p^{k+1}(x, p) = \partial F^{k+1}(x, p)/\partial p$ is invertible, i.e., $F^{k+1}(x, \dot{x}) = 0$ can be transformed to an ODE. For this recursive process to be applicable, a constant rank condition is required for the Jacobian $F_p^j(x, p)$ at each recursive step j. These results in [120], on characterizing the solvability of DAE systems through a derivation of the underlying ODE system, were further strengthened and expressed in terms of intrinsic geometric properties of the DAE system in [121].

In contrast with the above-mentioned results employing recursive definitions of constrained manifolds/mappings and imposing certain constant rank conditions in each recursive step, the approach of [18] involved a repetitive differentiation of Eq.1.15 to obtain sufficient conditions for solvability in terms of an extended set of equations called the derivative array equations. More specifically, differentiating Eq.1.15 k times with respect to t yields the following extended set of equations:

$$F_k(t, x, \dot{x}, w) = \begin{bmatrix} F(t, x, \dot{x}) \\ F_t(t, x, \dot{x}) + F_x(t, x, \dot{x})\dot{x} + F_{\dot{x}}(t, x, \dot{x})\ddot{x} \\ \vdots \\ \dfrac{d^k}{dt^k}[F(t, x, \dot{x}] \end{bmatrix} = 0 \qquad (1.17)$$

which involves higher-order time-derivatives of x, $w = [x^{(2)}, \ldots, x^{(k+1)}]$. While the above set of equations is still singular with respect to $x^{(k+1)}$ owing to the singularity of $F_{\dot{x}}(t, x, \dot{x})$, it is possible that they uniquely determine $\dot{x} = \phi(x, t)$ for some $k \geq 1$. Indeed, under certain rank conditions for suitable Jacobians of F_k, this is true and the DAE system is solvable. For a detailed description of these conditions and their comparison with other results on solvability conditions, see [18].

The common underlying principle in the approaches for characterizing the solvability of a DAE system is to obtain, either explicitly or implicitly, a local representation of an equivalent ODE system, for which available results on existence and uniqueness of solutions are applicable. This derivation of the underlying ODE system leads to the concept of index that is commonly used in the numerical simulation literature. For the general fully-implicit DAE system of 1.15, the index ν_d is defined as the smallest integer such that the derivative array equation $F_{\nu_d}(t, x, \dot{x}, w) = 0$ in Eq.1.17 uniquely determines $\dot{x}$ as a function of t, x [9]. In practice, depending on the structure of the system in Eq.1.15, it may not be necessary to differentiate all equations. In particular, for the semiexplicit DAE system of our interest in Eq.1.9, one needs to differentiate only the algebraic equations, and the index of such systems has the following definition [9].

Definition 1.1: The index ν_d of the DAE system in Eq.1.9 with specified smooth inputs $u(t)$, is the minimum number of times the algebraic equations or their subset have to be differentiated to obtain a set of differential equations for z, i.e., solve for $\dot{z} = \mathcal{F}(x, z, t)$.

The index ν_d provides a measure of the "singularity" of the algebraic equations and the resulting differences from ODE systems. More specifically, consider the DAE system of Eq.1.9 with a nonsingular matrix $l(x)$. Clearly, the algebraic equations can be solved for z:

$$z = -l(x)^{-1}[k(x) + c(x)u(t)] \qquad (1.18)$$

and one differentiation of the algebraic equations in Eq.1.9, or equivalently the solution for z, would yield the differential equations for z, i.e., the DAE system has an index $\nu_d = 1$. For such systems, a direct substitution of the solution for z in the differential equations for x, yields an equivalent ODE representation:

$$\dot{x} = \bar{f}(x) + \bar{g}(x)u(t) \qquad (1.19)$$

8

where

$$\bar{f}(x) = f(x) - b(x)l(x)^{-1}k(x)$$
$$\bar{g}(x) = g(x) - b(x)l(x)^{-1}c(x)$$

Thus, DAE systems of Eq.1.9 with an index one are essentially the same as ODE systems, and the simulation and control of such systems can be easily addressed on the basis of their ODE representation. Note that due to the linear appearance of the algebraic variables z in Eq.1.9, the linearity with respect to the inputs u is preserved in Eq.1.19.

In contrast, DAE systems with singular algebraic equations, more specifically a singular matrix $l(x)$, cannot be readily reduced into an ODE system and they have a high index $\nu_d > 1$. In this research note, we will focus on such high-index DAE systems. Moreover, we assume throughout that the DAE system in Eq.1.9 has a finite index for some smooth input $u(t)$, since this is necessary for the existence of a locally unique smooth solution $x(t), z(t)$.

Remark 1.2: Consider the following linear analogue of the semiexplicit DAE system in Eq.1.9:

$$\dot{x} = Ax + Bz + Gu(t)$$
$$0 = Kx + Lz + Cu(t)$$

(1.20)

where $x \in \mathbb{R}^n$, $z \in \mathbb{R}^p$, $u \in \mathbb{R}^m$, and A, B, G, K, L, and C are constant matrices of appropriate dimensions. The index ν_d of the above linear system is the same as the index of nilpotency of the matrix pencil:

$$\mathcal{P} = \begin{bmatrix} sI - A & -B \\ -K & -L \end{bmatrix}$$

(1.21)

Similarly, for the linear implicit system of Eq.1.1, the index ν_d is the same as the index of nilpotency of the matrix pencil $(sE - A)$ [9].

The index ν_d is also referred to as the differential (or differentiation) index. Another notion of index that has been used in the numerical simulation literature is the one of perturbation index. For the fully-implicit system of Eq.1.8, the perturbation index is defined [57] as the smallest integer ν_p such that the difference between the solution $x(t)$ and the solution $z(t)$ of the perturbed system:

$$F(\dot{z}, z, u) = e(t)$$

(1.22)

for the same inputs $u(t)$, can be bounded by an expression of the form:

$$\|z(t) - x(t)\| \leq K \left[\|z(0) - x(0)\| + \sum_{i=0}^{\nu_p - 1} \max_{0 \leq \tau \leq t} \left(\|e^{(i)}(\tau)\| \right) \right]$$

(1.23)

over a finite interval $t \in [0, T]$. Here $\|(\cdot)\|$ denotes a vector norm, $e^{(i)}(t)$ denotes the ith derivative of $e(t)$ and K is a scalar constant depending on the function F. The perturbation index ν_p provides a measure of the error in the numerical solution obtained in the presence of truncation and roundoff errors $(e(t) \neq 0)$. While for many DAE systems, the indices ν_d and ν_p satisfy $\nu_d \leq \nu_p \leq \nu_d + 1$ [49], they may be significantly different for general fully-implicit systems [17]. This is because these indices may vary significantly along the solution of Eq.1.22 with arbitrary perturbations $e(t)$. However, if one defines the maximum differential index $\nu_{d,max}$ and the maximum perturbation index $\nu_{p,max}$ with respect to the perturbation $e(t)$ over a suitable functional neighborhood of $e(t) = 0$, then the relation $\nu_{d,max} \leq \nu_{p,max} \leq \nu_{d,max} + 1$ holds for general implicit DAE systems [17]. The maximum differential index $\nu_{d,max}$ is also equal to a uniform differentiation index ν_{UD} defined on the basis of the derivative array equations in Eq.1.17 [17]. Another notion of index called structural index was also introduced in [132] on the basis of structural forms rather than the actual functional forms of the system equations. In general, the structural index ν_s provides only a lower bound for the differential index, i.e. $\nu_s \leq \nu_d$.

1.3.3   Comparison of high-index DAE and ODE systems

Consider an ODE system with the following description:

$$\dot{x} = f(x) + g(x)u(t) \tag{1.24}$$

where $x \in \mathcal{X} \subset \mathbb{R}^n$ is the vector of state variables, $u \in \mathbb{R}^m$ is the vector of input variables, $f(x)$ is smooth vector field, and $g(x)$ is a smooth matrix of dimension $n \times m$. The ODE system can be viewed as a special class of DAE systems of the form in Eq.1.9, with an index $\nu_d = 0$. As mentioned before, index-one DAE systems of Eq.1.9 with a nonsingular matrix $l(x)$ are also the same as ODE systems of Eq.1.24.

High-index DAE systems with a singular matrix $l(x)$ are, however, different from ODE systems. More specifically, the singular algebraic equations in Eq.1.9 imply the presence of underlying algebraic constraints in the differential variables x. Furthermore, if the index ν_d exceeds two, then the algebraic equations have to be differentiated several times (depending on ν_d) to obtain a solution for z, and the differentiated equations impose additional constraints in x. These constraints restrict the solution for x in a state space of dimension less than n. Consequently, for arbitrary initial conditions $x(0)$, the DAE system exhibits an impulsive behavior at the initial time $t = 0$. This is a common cause of failure in the numerical simulation of DAE systems. Only the initial conditions $x(0)$ that satisfy the underlying algebraic constraints in x, allow smooth solutions $(x(t), z(t))$ and are referred to as consistent initial conditions [15, 94, 110]. This is clearly in contrast with ODE systems of Eq.1.24, for which a well-defined solution exists for any initial condition $x(0)$. Furthermore, unlike ODE systems in Eq.1.24, the solution $(x(t), z(t))$ of high-index DAE systems in Eq.1.9 may also depend on the time-derivatives of the inputs $u(t)$, in which case these inputs must be sufficiently smooth. These fundamental differences between DAE and ODE systems are illustrated through the following simple example.

Example 1.1: Consider the following DAE system:

$$\dot{x}_1 = x_2 + z$$
$$\dot{x}_2 = x_1 + x_3$$
$$\dot{x}_3 = x_1 + x_2 + u(t)$$
$$0 = x_2 + x_3 \qquad (1.25)$$

with one algebraic variable z and one input $u(t)$ specified as a function of time. Note that the algebraic equation

$$0 = x_2 + x_3 \qquad (1.26)$$

does not involve z, i.e., $l(x) \equiv 0$, and thus, the DAE system has a high index. Moreover, the algebraic equation already denotes a constraint in the differential variables x. Differentiating this constraint once and using the relations for $\dot{x}_2$ and $\dot{x}_3$ yields the following new algebraic equation:

$$0 = 2x_1 + x_2 + x_3 + u(t) \qquad (1.27)$$

which denotes another constraint in x. Note however that this constraint explicitly involves the input $u(t)$. Another differentiation of this constraint yields the following equation:

$$0 = 2x_1 + 3x_2 + x_3 + u(t) + \dot{u}(t) + 2z \qquad (1.28)$$

which can be solved for z:

$$z = -0.5(2x_1 + 3x_2 + x_3 + u(t) + \dot{u}(t)) \qquad (1.29)$$

Thus, the DAE system in Eq.1.25 has an index $\nu_d = 3$; another differentiation of Eq.1.28 would yield the differential equation for z. The algebraic equations in Eq.1.26 and Eq.1.27 denote two constraints in x, one of which involves $u(t)$. These two constraints restrict the evolution of $x(t)$ in a one-dimensional state space, and a smooth solution $(x(t), z(t))$ exists only for consistent initial conditions $x(0)$ that satisfy the constraints $x_2(0) + x_3(0) = 0$, $2x_1(0) + x_2(0) + x_3(0) + u(0) = 0$. Note also that the solution for z depends on the derivative of the input, $\dot{u}(t)$, implying that the input $u(t)$ must be at least once continuously differentiable (as opposed to being just continuous for ODE systems).

The previous example illustrates the differences in the characteristics of ODE and high-index DAE systems, which arise due to the presence of underlying algebraic constraints in the differential variables x. The number and nature of these constraints clearly depend on the index ν_d. Note that the example is a linear DAE system of the form in Eq.1.20, for which the index ν_d is determined by the matrices A, B, K, L in the pencil $\mathcal{P}$ (Eq.1.21) and is completely independent of the inputs $u(t)$. However, in a nonlinear system of the form in Eq.1.9, the index ν_d itself may depend on the inputs $u(t)$. In fact, as illustrated by the following example, in the presence of arbitrarily time-varying inputs $u(t)$, a DAE system may even fail to have a well-defined index ν_d, leading to questions on solvability.

Example 1.2: Consider the DAE system:

$$
\begin{aligned}
\dot{x}_1 &= x_2 \\
\dot{x}_2 &= x_3 + x_4\, u(t) \\
\dot{x}_3 &= x_1 \\
\dot{x}_4 &= x_2 + z \\
0 &= x_1 + x_2
\end{aligned}
\tag{1.30}
$$

where again the algebraic equation does not involve z and denotes a constraint in x. Differentiating this constraint twice, the following two equations are obtained:

$$
0 = x_2 + x_3 + x_4\, u(t) \tag{1.31}
$$
$$
0 = x_1 + x_3 + (x_2 + x_4)u(t) + x_4\, \dot{u}(t) + u(t)\, z \tag{1.32}
$$

where Eq.1.31 denotes a constraint in the differential variables x, which also involves the input $u(t)$. Clearly, if $u(t) \neq 0$ on the time interval $[0, T]$, Eq.1.32 can be solved for z and the index is $\nu_d = 3$. However, if $u(t_1) = 0$ at any time $t_1 \in [0, T]$, then Eq.1.32 becomes singular and cannot be solved for z, indicating that at $t = t_1$, $\nu_d \neq 3$. Moreover, Eq.1.32 becomes another constraint in x implying a discontinuity in the dimension of the constrained state space where x evolves.

1.3.4 Regularity

The examples in the previous section clearly demonstrate that in a high-index DAE system of Eq.1.9, the underlying constraints in x may involve the inputs $u(t)$. In the context of numerical simulation of a DAE system with specified inputs $u(t)$, this implies that the inputs $u(t)$ must vary smoothly with time. However, in the context of feedback control, u is the vector of manipulated inputs that are not specified a *priori* as a function of time; rather we seek to design a feedback control law for these inputs. The presence of these manipulated inputs in the underlying algebraic constraints in x, has important ramifications on the controller design due to the fact that the constrained state-space region for x depends on the (unknown) feedback control law for u. The precise nature of the complications arising in the controller design from the presence of the inputs u in the underlying algebraic constraints, will become clear later in Chapters 3 and 4. For now, we simply present the following notion of regularity that is used in Chapters 3 and 4 to distinguish between two fundamentally different classes of regular and nonregular DAE systems and address the feedback control of these systems.

Definition 1.2: A DAE system of Eq.1.9 will be said to be regular, if

(i) it has a finite index ν_d, and

(ii) the state space region where the differential variables x are constrained to evolve is invariant under any control law for u.

Remark 1.3: Condition (ii) in the above definition of regular systems essentially states that the underlying algebraic constraints in x, which specify the constrained state space region, do not involve the inputs u. Thus, the requirement for a finite index in condition (i) is independent of the time-varying nature of these inputs, since the index does not depend on u under condition (ii).

The above notion of regularity is motivated in the context of feedback control of DAE systems, and it is different from existing notions of regular systems used in the literature in the context of solvability; in [9], a linear DAE system of Eq.1.1 is referred to as regular if the associated matrix pencil $\mathcal{P}$ is regular, while in [122, 123], a nonlinear DAE system is said to be regular if it corresponds to a locally unique vector field on a manifold. Clearly, on the basis of the representation of the DAE system in Eq.1.9, it is not obvious whether the system is regular or not. A procedure for the classification of a DAE system of Eq.1.9 as regular/nonregular will be developed in Chapter 3. Moreover, in Chapter 3 we will focus on the control of regular systems, while the control of nonregular systems will be addressed in Chapter 4.

2. Chemical Process Applications

Chemical processes are quite naturally modeled by systems of coupled differential and algebraic equations, where the differential equations arise from the dynamic balances of mass, energy, and momentum, while the algebraic equations typically include thermodynamic relations, empirical correlations, and quasi-steady-state relations. Often, the algebraic equations are nonsingular and can be readily eliminated from the DAE models to obtain standard ODE models. This is evident in the abundant literature on the modeling and control of chemical processes, which has focused almost exclusively on ODE systems. The occurrence of high-index DAE models was documented only recently in [46, 93, 111]. However, the source of high index in these DAE models has often been ascribed to "improper modeling" and "incorrect choice" of process specifications.

In this chapter, we present a few examples of chemical processes that are modeled by high-index DAEs. Through these examples, we illustrate that high-index DAE models arise naturally in a wide variety of chemical process applications with simultaneous fast and slow phenomena, e.g., reactors with fast and slow reactions or fast heat/mass transfer coupled with slow reactions. For these processes, detailed "rate-based" models where the rate expressions for the fast phenomena are explicitly included in the modeling equations, are typically given by index-one DAE or ODE models. However, such models exhibit stiffness/time-scale multiplicity, which leads to well-known problems in numerical simulation; these models are not particularly well-suited for controller design either. On the other hand, when these fast phenomena are approximated with appropriate quasi-steady-state (QSS) conditions, the resulting "equilibrium-based" models are given by high-index DAEs, which describe the essential slow dynamics of the process and are better suited for simulation and controller design purposes.

In some processes the fast dynamics correspond to specific physical phenomena like fast mass transfer/reactions and the QSS conditions are intuitively apparent (e.g., phase/reaction equilibrium), whereas in other processes, the QSS conditions are not so apparent. However, all these processes share the characteristic feature that the detailed ODE models of these processes involve large parameters (e.g., large mass transfer and reaction rate coefficients for fast mass transfer/reactions), which lead to the stiffness/time-scale multiplicity. We shall discuss this in more detail in Chapter 6 where we analyze the dynamic behavior of ODE systems with large parameters within the framework of singular perturbations, and provide a rigorous justification for the QSS approximations and the high-index DAE models. For now, we simply present the high-index DAE models obtained under the appropriate QSS assumptions.

2.1 Reactor with Fast Heat Transfer through Heating Jacket

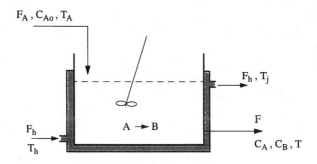

Figure 2.1: A CSTR with heating jacket

Consider the continuously stirred tank reactor (CSTR) with heating jacket in Figure 2.1. Reactant A is fed at a flow rate F_A, molar concentration C_{Ao}, and temperature T_A to the reactor, where the irreversible endothermic reaction $A \to B$ occurs. The rate of reaction is given by the following relation:

$$R_A = k_o \exp(-E_a/RT)C_A$$

where C_A is the molar concentration of A in the reactor holdup, and T is the reactor temperature. The product stream is withdrawn at a flow rate F and heat is provided to the reactor through the heating jacket, where the heating fluid is fed at a flow rate F_h and temperature T_h. Consider the case when the heat transfer rate $Q = UA(T_j - T)$ is fast, i.e., the product of overall heat transfer coefficient U and heat transfer area A is large. The detailed rate-based model of the process, where the heat transfer rate expression is explicitly included, is given by the following ODE system:

$$\dot{V} = F_A - F$$

$$\dot{C}_A = \frac{F_A}{V}(C_{Ao} - C_A) - k_o \exp(-E_a/RT)C_A$$

$$\dot{C}_B = -\frac{F_A}{V}C_B + k_o \exp(-E_a/RT)C_A$$

$$\dot{T} = \frac{F_A}{V}(T_A - T) - k_o \exp(-E_a/RT)C_A \frac{\Delta H_r}{\rho c_p} + \frac{Q}{\rho V c_p}$$

$$\dot{T}_j = \frac{F_h}{V_h}(T_h - T_j) - \frac{Q}{\rho_h V_h c_{ph}}$$

$$0 = Q - UA(T_j - T) \tag{2.1}$$

15

Owing to the presence of the large parameter UA in the rate expression for the fast heat transfer, the ODE model in Eq.2.1 exhibits stiffness. Equivalently, the fast heat transfer implies a time-scale multiplicity in the process dynamics where after an initial fast transience, the reactor and the jacket are essentially at thermal equilibrium, i.e., $T_j \approx T$. Thus, under the QSS assumption of thermal equilibrium, the explicit rate expression for the fast heat transfer is replaced by the relation $T_j = T$, to obtain the following DAE model:

$$\dot{V} = F_A - F$$
$$\dot{C}_A = \frac{F_A}{V}(C_{Ao} - C_A) - k_o \exp(-E_a/RT)C_A$$
$$\dot{C}_B = -\frac{F_A}{V}C_B + k_o \exp(-E_a/RT)C_A$$
$$\dot{T} = \frac{F_A}{V}(T_A - T) - k_o \exp(-E_a/RT)C_A \frac{\Delta H_r}{\rho c_p} + \frac{Q}{\rho V c_p}$$
$$\dot{T}_j = \frac{F_h}{V_h}(T_h - T_j) - \frac{Q}{\rho_h V_h c_{ph}} \qquad (2.2)$$
$$0 = T_j - T$$

Clearly, due to the assumption of thermal equilibrium, the algebraic equation are singular and can not be solved for the algebraic variable Q. It can be verified that the above DAE model has an index two.

2.2   Reactor with Fast and Slow Reactions

Consider an isothermal CSTR where reactant A is fed at a flow rate F_o and concentration C_{Ao}, and the following elementary reactions occur in series:

$$A \rightleftharpoons B \rightarrow C$$

with the net forward rates of reactions R_1 and R_2, respectively, given by the following:

$$R_1 = k_1 \left(C_A - \frac{C_B}{K_{eq}} \right) , \qquad R_2 = k_2 C_B$$

The reversible reaction $A \rightleftharpoons B$ is much faster than the irreversible reaction $B \rightarrow C$, i.e., $k_1 \gg k_2$. A detailed rate-based model, including the rate expressions for both reactions, is given by the following ODE model:

$$\dot{V} = F_o - F$$
$$\dot{C}_A = \frac{F_o}{V}(C_{Ao} - C_A) - R_1$$

$$\dot{C}_B = -\frac{F_o}{V}C_B + R_1 - R_2$$

$$\dot{C}_C = -\frac{F_o}{V}C_C + R_2 \qquad (2.3)$$

$$0 = R_1 - k_1 \left(C_A - \frac{C_B}{K_{eq}} \right)$$

$$0 = R_2 - k_2 C_B$$

Clearly, the algebraic equations can be solved for the reaction rates R_1 and R_2, and substituting these relations in the differential equations, the above index-one DAE model is readily reduced to an ODE model. However, the above model involves the large reaction rate coefficient k_1 of the fast reversible reaction and consequently exhibits stiffness.

The fact that the first reversible reaction is much faster than the second irreversible one, implies that the reversible reaction is essentially at equilibrium after an initial fast transience. Under this QSS assumption of reaction equilibrium for the fast reversible reaction, the following DAE model is obtained:

$$\dot{V} = F_o - F$$

$$\dot{C}_A = \frac{F_o}{V}(C_{Ao} - C_A) - R_1$$

$$\dot{C}_B = -\frac{F_o}{V}C_B + R_1 - R_2$$

$$\dot{C}_C = -\frac{F_o}{V}C_C + R_2 \qquad (2.4)$$

$$0 = C_A - \frac{C_B}{K_{eq}}$$

$$0 = R_2 - k_2 C_B$$

Clearly, while the second algebraic equation yields the explicit relation for the slow reaction rate R_2, the reaction rate R_1 for the fast reversible reaction does not appear in any of the algebraic equations, i.e., the algebraic equations are singular. The DAE model in Eq.2.4 has an index two.

In many practical applications, e.g., chemical vapor deposition (CVD), catalytic crackers, and combustion, several reactions occur simultaneously, some of which are often much faster than the others. The above example illustrates that the QSS assumptions of reaction equilibrium for fast reversible reactions, and similarly complete conversion for fast irreversible reactions, lead to high-index DAE models of such processes.

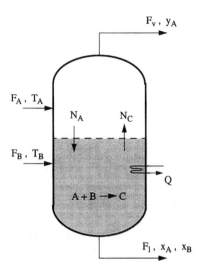

F_v, y_A

F_A, T_A

N_A N_C

F_B, T_B

Q

$A + B \rightarrow C$

F_l, x_A, x_B

Figure 2.2: Two-phase reactor

2.3   Two-Phase Reactor with Fast Mass Transfer

Consider the two-phase (liquid/vapor) reactor in Figure 2.2, where reactants A and B are fed as pure vapor and liquid streams at molar flow rates F_A and F_B and temperatures T_A and T_B, respectively. The reactant A diffuses into the liquid phase at a rate N_A, where the exothermic irreversible reaction

$$A + B \rightarrow C$$

yields the product C, which diffuses into the vapor phase at a rate N_C; reactant B is nonvolatile compared to A and C. The inter-phase mass transfer of A and C is much faster than the liquid-phase reaction. The mass transfer rates are given by the correlations:

$$N_A = k_A a (y_A - y_A^*) \frac{M_l}{\rho}$$
$$N_C = k_C a (y_C^* - (1 - y_A)) \frac{M_l}{\rho} \qquad (2.5)$$

where k_A, k_C are the overall mass transfer coefficients, a is the mass transfer area per unit liquid holdup volume, y_A is the mole fraction of A in vapor phase, y_A^*, y_C^* denote equilibrium vapor phase mole fractions corresponding to the liquid phase composition, and M_l, ρ are the molar liquid holdup and density. The rate of production of C

through the reaction is given by the relation:

$$R_C = k_o \exp(-E_a/RT) M_l \, \rho \, x_A x_B \qquad (2.6)$$

where T is the reactor temperature, and x_A and x_B are the liquid-phase mole fractions of A and B.

The product vapor and liquid streams are withdrawn from the reactor at molar flow rates F_v and F_l, respectively. It is assumed that all components have equal and constant molar density ρ, heat capacity c_p, and heat of vaporization ΔH^v, and the liquid and vapor phases are well-mixed and behave ideally. Under these assumptions, the following DAE model is obtained for the process:

$$\dot{M}_v = F_A - N_A + N_C - F_v$$

$$\dot{y}_A = \frac{1}{M_v}[F_A(1 - y_A) - N_A(1 - y_A) - N_C y_A]$$

$$\dot{M}_l = F_B - F_l - R_C + N_A - N_C$$

$$\dot{x}_A = \frac{1}{M_l}[-F_B x_A - R_C(1 - x_A) + N_A(1 - x_A) + N_C x_A]$$

$$\dot{x}_B = \frac{1}{M_l}[F_B(1 - x_B) - R_C(1 - x_B) - N_A x_B + N_C x_B]$$

$$\dot{T} = \frac{1}{(M_l + M_v)c_p}[F_A c_p(T_A - T) + F_B c_p(T_B - T) + (N_A - N_C)\Delta H^v$$
$$-R_C(\Delta H_R^o - c_p(T - T_o)) - Q]$$

$$0 = N_A - k_A a(y_A - y_A^*)\frac{M_l}{\rho}$$

$$0 = N_C - k_C a(y_C^* - (1 - y_A))\frac{M_l}{\rho}$$

$$0 = P_A^s x_A - P y_A^* \qquad (2.7)$$

$$0 = P_C^s(1 - x_A - x_B) - P y_C^*$$

$$0 = M_v RT - P(V_T - \frac{M_l}{\rho})$$

where M_v is the molar vapor holdup, V_T is the total reactor volume, P is the reactor pressure and P_A^s, P_C^s denote the saturation vapor pressures of A and C given by the standard Antoine relations:

$$P_A^s = \exp\left(A_A - \frac{B_A}{T + C_A}\right)$$

$$P_C^s = \exp\left(A_C - \frac{B_C}{T + C_C}\right)$$

It is straightforward to verify that the above DAE model has an index one and can be reduced to an ODE model by substituting the relations for the inter-phase mass transfer rates N_A and N_C in the differential equations. However, this model exhibits stiffness due to the presence of the large mass transfer coefficients in the mass transfer correlations.

The fast inter-phase mass transfer implies that after a fast initial transience, the liquid and vapor phase compositions are close to equilibrium. Thus, under the QSS approximation of phase equilibrium, the explicit mass transfer correlations in Eq.2.7 are replaced by the phase equilibrium relations $(y_A = y_A^*, \ y_C = 1 - y_A = y_C^*)$ to obtain the following DAE model:

$$
\begin{aligned}
\dot{M}_v &= F_A - N_A + N_C - F_v \\
\dot{y}_A &= \frac{1}{M_v}[F_A(1 - y_A) - N_A(1 - y_A) - N_C y_A] \\
\dot{M}_l &= F_B - F_l - R_C + N_A - N_C \\
\dot{x}_A &= \frac{1}{M_l}[-F_B x_A - R_C(1 - x_A) + N_A(1 - x_A) + N_C x_A] \\
\dot{x}_B &= \frac{1}{M_l}[F_B(1 - x_B) - R_C(1 - x_B) - N_A x_B + N_C x_B] \\
\dot{T} &= \frac{1}{(M_l + M_v)c_p}[F_A c_p(T_A - T) + F_B c_p(T_B - T) + (N_A - N_C)\Delta H^v \\
&\qquad\qquad - R_C(\Delta H_R^o - c_p(T - T_o)) - Q] \\
0 &= P_A^s x_A - P y_A \\
0 &= P_C^s(1 - x_A - x_B) - P(1 - y_A) \\
0 &= M_v RT - P(V_T - \frac{M_l}{\rho})
\end{aligned}
\tag{2.8}
$$

Clearly, due to the assumption of phase equilibrium for the fast mass transfer, the inter-phase mass transfer rates N_A and N_C do not appear in the algebraic equations. Thus, the above equilibrium-based DAE model has a high-index; it can be verified that the index is two.

A conventional approach to modeling such vapor-liquid multi-phase processes is to assume that the vapor holdup M_v is negligible compared to the liquid holdup M_l, to obtain the following DAE model:

$$
\begin{aligned}
\dot{M}_l &= F_A + F_B - F_l - F_v - R_C \\
\dot{x}_A &= \frac{1}{M_l}[F_A(1 - x_A) - F_B x_A - F_v(y_A - x_A) - R_C(1 - x_A)] \\
\dot{x}_B &= \frac{1}{M_l}[-F_A x_B + F_B(1 - x_B) + F_v x_B - R_C(1 - x_B)] \\
\dot{T} &= \frac{1}{M_l c_p}[F_A(c_p(T_A - T) + \Delta H^v) + F_B c_p(T_B - T) - F_v \Delta H^v \\
&\qquad\qquad - R_C(\Delta H_R^o - c_p(T - T_o)) - Q]
\end{aligned}
\tag{2.9}
$$

$$0 = P_A^s x_A - P y_A$$
$$0 = P_C^s(1 - x_A - x_B) - P(1 - y_A)$$

which does not include the dynamic balances for the vapor holdup, and consequently does not involve the inter-phase mass transfer rates N_A or N_C. The above simplified model has index one and is easily reduced to an ODE model by solving for the vapor phase mole fraction y_A from the algebraic equations and substituting the solution in the differential equations.

The assumption of negligible vapor holdup is valid only at low operating pressures. Moreover, even at low or moderate pressure, this assumption implies that *all* the gaseous reactant A is fed directly to the liquid phase, which is the reaction phase, without any mass transfer (compare the equations for $\dot{x}_A$ in Eq.2.9 and Eq.2.8). This is clearly not true, and the index-one model in Eq.2.9 may predict an erroneous dynamic behavior of the process (see Section 7.3.3 for a comparison of the dynamic behavior predicted by the models in Eq.2.8 and Eq.2.9). The extent of this error depends, in general, on factors like the relative volatilities of reactants and products, and whether the reaction is endothermic or exothermic. Similar modeling issues arise in reactive distillation columns, where each tray is like a two-phase reactor discussed above, and the vapor phase may have a significant role in the coupled reaction-separation dynamics [86].

2.4 Cascade of Reactors with High Pressure Gaseous Flow

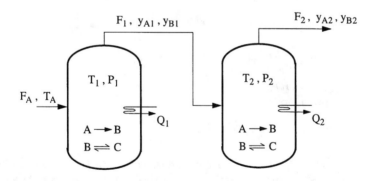

Figure 2.3: Cascade of two gas phase reactors

Consider the cascade of two gas phase CSTRs in Figure 2.3. Gaseous reactant A is fed to the first reactor at a molar flow rate F_A and temperature T_A, where the exothermic reactions $A \rightarrow B$, $B \rightleftharpoons C$ occur in series. The reactor operates at a high temperature T_1 to promote the conversion of A to B in the first irreversible reaction. However, the production of C is restricted by the thermodynamic equilibrium limitation for the reversible reaction. The gaseous mixture from the first reactor is fed to the second reactor, which operates at a lower temperature T_2 to shift the reaction equilibrium in the forward direction, favoring the production of C. The product gaseous stream from the second reactor is withdrawn at a molar flow rate F_2. On the other hand, the molar flow rate F_1 of the gaseous stream from the first to the second reactor is governed by the pressure drop $\Delta P = P_1 - P_2$ between the two reactors:

$$F_1 = (\frac{P_1}{RT_1})\frac{(\Delta P)^{4/7}}{\sigma}$$

The overall model of the reactor cascade is given by the following DAE system:

$$\dot{M}_1 = F_A - F_1$$
$$\dot{y}_{A1} = \frac{1}{M_1}[F_A(1 - y_{A1}) - R_{B,1}]$$
$$\dot{y}_{B1} = \frac{1}{M_1}[-F_A y_{B1} + R_{B,1} - R_{C,1}]$$
$$\dot{T}_1 = \frac{1}{M_1 c_p}[F_A c_p(T_A - T_1) - R_{B,1}\Delta H_{R1} - R_{C,1}\Delta H_{R2} - Q_1]$$
$$\dot{M}_2 = F_1 - F_2$$
$$\dot{y}_{A2} = \frac{1}{M_2}[F_1(y_{A1} - y_{A2}) - R_{B,2}]$$
$$\dot{y}_{B2} = \frac{1}{M_2}[F_1(y_{B1} - y_{B2}) + R_{B,2} - R_{C,2}]$$
$$\dot{T}_2 = \frac{1}{M_2 c_p}[F_1 c_p(T_1 - T_2) - R_{B,2}\Delta H_{R1} - R_{C,2}\Delta H_{R2} - Q_2]$$
$$0 = M_1 RT_1 - P_1 V_{1T}$$
$$0 = M_2 RT_2 - P_2 V_{2T}$$
$$0 = F_1 - (\frac{P_1}{RT_1})\frac{(P_1 - P_2)^{4/7}}{\sigma}$$

$$(2.10)$$

In the above model, $R_{B,i}$ and $R_{C,i}$ denote the rate of production of B and C through the irreversible and reversible reactions, respectively, in the first ($i = 1$) and second ($i = 2$) reactors, given by:

$$R_{B,i} = k_{1,o}\exp(-E_1/RT_i)M_i y_{Ai}$$
$$R_{C,i} = k_{2,o}\exp(-E_2/RT_i)M_i \left(y_{Bi} - \frac{y_{Ci}}{K_{eq}(T_i)}\right), \quad i = 1,2$$

Clearly, the algebraic equations in Eq 2.10 can be solved for P_1, P_2, and F_1, i.e., the DAE model has an index one and can be reduced to an ODE system by substituting the solution for F_1 in the differential equations.

However, at high operating pressures the molar density of the gas, P_1/RT_1, is high. Owing to this high gas density, even a small pressure drop ΔP yields a high molar flow rate F_1, and the index-one DAE model in Eq.2.10 exhibits stiffness. Equivalently, the process has a fast dynamics associated with the "fast" gaseous flow F_1, and after the initial transience, the pressure P_1 reaches close to P_2. Thus, under the QSS assumption of negligible pressure drop, the explicit correlation for the gas flow rate F_1 is replaced by $P_1 = P_2$ to obtain the following DAE model:

$$\dot{M}_1 = F_A - F_1$$
$$\dot{y}_{A1} = \frac{1}{M_1}[F_A(1 - y_{A1}) - R_{B,1}]$$
$$\dot{y}_{B1} = \frac{1}{M_1}[-F_A y_{B1} + R_{B,1} - R_{C,1}]$$
$$\dot{T}_1 = \frac{1}{M_1 c_p}[F_A c_p(T_A - T_1) - R_{B,1}\Delta H_{R1} - R_{C,1}\Delta H_{R2} - Q_1]$$
$$\dot{M}_2 = F_1 - F_2$$
$$\dot{y}_{A2} = \frac{1}{M_2}[F_1(y_{A1} - y_{A2}) - R_{B,2}]$$
$$\dot{y}_{B2} = \frac{1}{M_2}[F_1(y_{B1} - y_{B2}) + R_{B,2} - R_{C,2}]$$
$$\dot{T}_2 = \frac{1}{M_2 c_p}[F_1 c_p(T_1 - T_2) - R_{B,2}\Delta H_{R1} - R_{C,2}\Delta H_{R2} - Q_2]$$
$$0 = M_1 RT_1 - P_1 V_{1T}$$
$$0 = M_2 RT_2 - P_2 V_{2T} \tag{2.11}$$
$$0 = P_1 - P_2$$

The gas flow rate F_1 clearly does not appear in any of the algebraic equations, i.e., the algebraic equations are singular, and the DAE model in Eq.2.11 has an index two. The ideal gas equations for the holdups in the two reactors of fixed volumes V_{1T}, V_{2T}, respectively, are used for simplicity, and could be replaced by a nonideal equation of state without affecting this conclusion. Such high-index DAE models arise in many staged processes with gaseous flow at high pressures, e.g., high-purity absorption, distillation, or reactive distillation columns, where the column pressure is high and the pressure drop between successive stages is small.

2.5 Reaction-Separation Process Network with Large Recycle

In the previous examples, we considered isolated reaction/separation units where fast reactions, mass transfer, etc., caused the time-scale multiplicity in the process dynamics. A typical industrial process however consists of a network of reactors and separators with recycle. In this section we present a simple example of such a reaction-separation network, where the recycle flow rate is large compared to the throughput. Consequently, the detailed model of the overall process network exhibits stiffness, and a model for the slow dynamics of the network obtained under appropriate QSS conditions, which are not as apparent as in the previous examples, is given by a high-index DAE system.

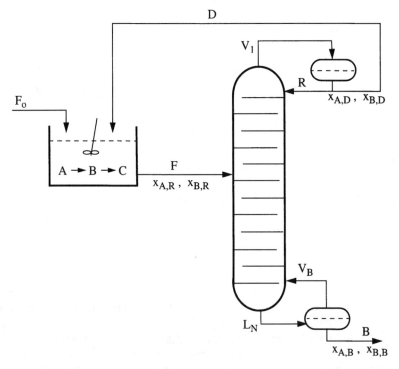

Figure 2.4: A network of CSTR and distillation column with recycle

Consider the network of a CSTR and distillation column in Figure 2.4. Reactant A is fed at a molar flow rate F_o to the CSTR, where the first-order irreversible reactions $A \rightarrow B \rightarrow C$ produce the desired product B and the undesired byproduct C. The outlet stream from the reactor, which is a mixture of the unconverted reactant A

and the products B and C, is fed to the distillation column with N trays (numbered from top to bottom) on tray f at a flow rate F. The light unconverted reactant A is distilled at the top of the column and recycled completely to the reactor at a flow rate D, while the heavier product B and byproduct C are withdrawn at the bottom from the reboiler at a flow rate B. For simplicity, we assume isothermal operation of the reactor, constant molar overflow and relative volatilities $\alpha_A > \alpha_B > \alpha_C = 1$, and equal latent heat of vaporization for all components. Under these assumptions, the vapor flow rate V_i leaving each tray i is equal to the vapor boilup V_B from the reboiler, while the liquid flow rate L_i is equal to the liquid recycle R in the enriching section and $R + F$ in the stripping section. A detailed dynamic model for the reactor-distillation column network is given by the following ODE system:

$$
\left.
\begin{aligned}
\dot{M}_R &= F_o + D - F \\
\dot{x}_{A,R} &= \frac{F_o(1 - x_{A,R}) + D(x_{A,D} - x_{A,R})}{M_R} - k_1 x_{A,R} \\
\dot{x}_{B,R} &= \frac{-F_o x_{B,R} + D(x_{B,D} - x_{B,R})}{M_R} + k_1 x_{A,R} - k_2 x_{B,R}
\end{aligned}
\right\} \text{ reactor}
$$

$$
\left.
\begin{aligned}
\dot{M}_D &= V_B - R - D \\
\dot{x}_{A,D} &= \frac{V_B(y_{A,1} - x_{A,D})}{M_D} \\
\dot{x}_{B,D} &= \frac{V_B(y_{B,1} - x_{B,D})}{M_D}
\end{aligned}
\right\} \text{ condenser}
$$

$$
\left.
\begin{aligned}
\dot{x}_{A,i} &= \frac{1}{M_i}[V_B(y_{A,i+1} - y_{A,i}) + R(x_{A,i-1} - x_{A,i})] \\
\dot{x}_{B,i} &= \frac{1}{M_i}[V_B(y_{B,i+1} - y_{B,i}) + R(x_{B,i-1} - x_{B,i})]
\end{aligned}
\right\} \text{ tray } i, \; 1 \leq i < f
$$

$$
\left.
\begin{aligned}
\dot{x}_{A,f} &= \frac{1}{M_f}[V_B(y_{A,f+1} - y_{A,f}) + R(x_{A,f-1} - x_{A,f}) + F(x_{A,R} - x_{A,f})] \\
\dot{x}_{B,f} &= \frac{1}{M_f}[V_B(y_{B,f+1} - y_{B,f}) + R(x_{B,f-1} - x_{B,f}) + F(x_{B,R} - x_{B,f})]
\end{aligned}
\right\} \text{ tray } f
$$

$$
\left.
\begin{aligned}
\dot{x}_{A,i} &= \frac{1}{M_i}[V_B(y_{A,i+1} - y_{A,i}) + (R+F)(x_{A,i-1} - x_{A,i})] \\
\dot{x}_{B,i} &= \frac{1}{M_i}[V_B(y_{B,i+1} - y_{B,i}) + (R+F)(x_{B,i-1} - x_{B,i})]
\end{aligned}
\right\} \text{ tray } i, \; f < i \leq N
$$

$$
\left.
\begin{aligned}
\dot{M}_B &= R + F - V_B - B \\
\dot{x}_{A,B} &= \frac{1}{M_B}[(R+F)(x_{A,N} - x_{A,B}) - V_B(y_{A,B} - x_{A,B})] \\
\dot{x}_{B,B} &= \frac{1}{M_B}[(R+F)(x_{B,N} - x_{B,B}) - V_B(y_{B,B} - x_{B,B})]
\end{aligned}
\right\} \text{ reboiler} \qquad (2.12)
$$

where M_R, M_D, M_B, M_i denote the molar liquid holdups in the reactor, condenser, reboiler, and trays i, $x_{A,i}, x_{B,i}$, etc., denote the corresponding mole fractions of A and B, and $y_{A,i}, y_{B,i}$ denote the vapor phase mole fractions given by the following relations:

$$y_{A,i} = \frac{\alpha_A x_{A,i}}{1 + (\alpha_A - 1)x_{A,i} + (\alpha_B - 1)x_{B,i}}, \quad y_{B,i} = \frac{\alpha_B x_{B,i}}{1 + (\alpha_A - 1)x_{A,i} + (\alpha_B - 1)x_{B,i}}$$

In this process, the two reactions with rate constants $k_1 \approx k_2$ compete in series to produce the main product B and byproduct C, and it is desired to have high conversion of the reactant A and a high product selectivity for B. This is achieved by keeping the single-pass conversion low in the reactor, and using a recycle flow rate D much larger than the reactant feed flow rate F_o or product flow rate B. This large recycle flow rate compared to the throughput introduces a stiffness or time-scale multiplicity in the detailed model of Eq.2.12 for the overall process network. A model for the slow process dynamics is obtained under appropriate QSS approximations for the fast dynamics.

The QSS conditions for the above process network are not as intuitively apparent as the QSS conditions of phase, thermal, reaction equilibrium, etc., corresponding to fast mass/heat transfer and fast reactions in the previous examples. However, they are associated with terms involving large parameters, e.g., the large recycle flow rate D, very much like the rate expressions of fast reactions/mass transfer involving large rate coefficients (for a more detailed and rigorous discussion, see Section 6.4). This observation suggests isolating the terms with large parameters to identify the QSS conditions for the fast dynamics. Note first that a large recycle flow rate D implies that the flow rates F and V_B are equally large. Thus, introducing the variables $\kappa_1 = D/D_{nom}$, $\kappa_2 = F/D_{nom}$ and $\kappa_3 = V_B/D_{nom}$ where D_{nom} refers to the nominal value of D (these flow rates are typically variable) and $\kappa_1 \approx \kappa_2 \approx \kappa_3$, the following terms involving the large flow rate D_{nom} in the differential equations of Eq.2.12 can be isolated:

$$z_R^0 = D_{nom}(\kappa_1 - \kappa_2), \quad z_R^1 = D_{nom}\kappa_1(x_{A,D} - x_{A,R}), \quad z_R^2 = D_{nom}\kappa_1(x_{B,D} - x_{B,R})$$
$$z_D^0 = D_{nom}(\kappa_3 - \kappa_1), \quad z_D^1 = D_{nom}\kappa_3(y_{A,1} - x_{A,D}), \quad z_D^2 = D_{nom}\kappa_3(y_{B,1} - x_{B,D})$$
$$z_i^1 = D_{nom}\kappa_3(y_{A,i+1} - y_{A,i}), \quad z_i^2 = D_{nom}\kappa_3(y_{B,i+1} - y_{B,i}) \quad i = 1, \ldots, f-1$$
$$z_f^1 = D_{nom}[\kappa_3(y_{A,f+1} - y_{A,f}) + \kappa_2(x_{A,R} - x_{A,f})]$$
$$z_f^2 = D_{nom}[\kappa_3(y_{B,f+1} - y_{B,f}) + \kappa_2(x_{B,R} - x_{B,f})] \tag{2.13}$$
$$\left. \begin{array}{l} z_i^1 = D_{nom}[\kappa_3(y_{A,i+1} - y_{A,i}) + \kappa_2(x_{A,i-1} - x_{A,i})] \\ z_i^2 = D_{nom}[\kappa_3(y_{B,i+1} - y_{B,i}) + \kappa_2(x_{B,i-1} - x_{B,i})] \end{array} \right\} \quad i = f+1, \ldots, N$$

In the limit when the large parameter $D_{nom} \to \infty$ in the above terms, the remaining factors approach zero, thus leading to the following QSS conditions:

$$0 = \kappa_1 - \kappa_2, \quad 0 = \kappa_1(x_{A,D} - x_{A,R}), \quad 0 = \kappa_1(x_{B,D} - x_{B,R})$$
$$0 = \kappa_3 - \kappa_1, \quad 0 = \kappa_3(y_{A,1} - x_{A,D}), \quad 0 = \kappa_3(y_{B,1} - x_{B,D})$$

$$0 = \kappa_3(y_{A,i+1} - y_{A,i}), \quad 0 = \kappa_3(y_{B,i+1} - y_{B,i}) \quad i = 1, \ldots, f-1$$

$$0 = \kappa_3(y_{A,f+1} - y_{A,f}) + \kappa_2(x_{A,R} - x_{A,f})$$

$$0 = \kappa_3(y_{B,f+1} - y_{B,f}) + \kappa_2(x_{B,R} - x_{B,f}) \tag{2.14}$$

$$\left.\begin{array}{l} 0 = \kappa_3(y_{A,i+1} - y_{A,i}) + \kappa_2(x_{A,i-1} - x_{A,i}) \\ 0 = \kappa_3(y_{B,i+1} - y_{B,i}) + \kappa_2(x_{B,i-1} - x_{B,i}) \end{array}\right\} \quad i = f+1, \ldots, N$$

Note that in this limit $D_{nom} \to \infty$, the variables defined in Eq.2.13 become indeterminate, analogous to the indeterminate fast mass transfer/reaction rates. Thus, these variables are the unknown algebraic variables z that vary according to the QSS relations in Eq.2.14. The identification of the QSS conditions and the definition of the algebraic variables yield a DAE model for the slow dynamics of the overall network, which consists of the algebraic equations in Eq.2.14 and the following differential equations:

$$\left.\begin{array}{l} \dot{M}_R = F_o + z_R^0 \\[2mm] \dot{x}_{A,R} = \dfrac{F_o(1 - x_{A,R}) + z_R^1}{M_R} - k_1 x_{A,R} \\[2mm] \dot{x}_{B,R} = \dfrac{-F_o x_{B,R} + z_R^2}{M_R} + k_1 x_{A,R} - k_2 x_{B,R} \end{array}\right\} \text{ reactor}$$

$$\left.\begin{array}{l} \dot{M}_D = z_D^0 - R \\[2mm] \dot{x}_{A,D} = \dfrac{z_D^1}{M_D} \\[2mm] \dot{x}_{B,D} = \dfrac{z_D^2}{M_D} \end{array}\right\} \text{ condenser}$$

$$\left.\begin{array}{l} \dot{x}_{A,i} = \dfrac{1}{M_i}[z_i^1 + R(x_{A,i-1} - x_{A,i})] \\[2mm] \dot{x}_{B,i} = \dfrac{1}{M_i}[z_i^2 + R(x_{B,i-1} - x_{B,i})] \end{array}\right\} \text{ tray } i, \ 1 \le i \le N$$

$$\left.\begin{array}{l} \dot{M}_B = R - B - (z_R^0 + z_D^0) \\[2mm] \dot{x}_{A,B} = \dfrac{1}{M_B}[R(x_{A,N} - x_{A,B}) + \gamma_1(z)] \\[2mm] \dot{x}_{B,B} = \dfrac{1}{M_B}[R(x_{B,N} - x_{B,B}) + \gamma_2(z)] \end{array}\right\} \text{ reboiler} \tag{2.15}$$

In the above DAE system $\gamma_1(z)$, $\gamma_2(z)$ denote certain linear combinations of the algebraic variables z defined in Eq.2.13. The exact form of these terms follows from expressing the constraints:

$$0 = \kappa_2 - \kappa_3$$

$$0 = \kappa_2(x_{A,N} - x_{A,B}) - \kappa_3(y_{A,B} - x_{A,B})$$

$$0 = \kappa_2(x_{B,N} - x_{B,B}) - \kappa_3(y_{B,B} - x_{B,B})$$

as a linear combination of the constraints in Eq.2.14, and is omitted for brevity.

Clearly, the above DAE model for the slow process dynamics has a high index since the algebraic variables do not appear in any of the algebraic equations in Eq.2.14. Moreover, the algebraic constraints in Eq.2.14 involve the large flow rates D, F, and V_B, or equivalently κ_i, which typically are time-varying inputs in the process. Thus, unlike the previous examples, the index of the DAE model for this example depends on the specific nature of these inputs and the solution of the DAE system depends on the time-derivatives of these inputs (see Section 1.3.3). In fact, the DAE model does not have a well-defined index for the case when these large flow rates are specified as functions of time; a well-defined index exists only if these inputs vary as function of the differential variables, i.e., according to some feedback law. An interesting consequence of the constraints in Eq.2.14 is that $x_{A,i} = x_{A,i+1}$ and $x_{B,i} = x_{B,i+1}$ in the enriching section of the column, i.e., most of the distillation occurs only in the stripping section, and the reactor outlet should be fed to the top tray to achieve maximum separation of the unconverted reactant A from the products B and C.

2.6 Reactor with High-Gain Pressure Controller

Unlike the previous examples where the fast dynamics were associated with some inherent physical phenomena in the process, fast dynamics may also be *induced* by a high-gain controller. The detailed dynamic models for such processes, where the high-gain controller equation is explicitly included in the modeling equations, exhibit a similar stiffness/time-scale multiplicity owing to the presence of the large controller gain. When the fast dynamics arising from the high-gain controllers are approximated with the corresponding QSS conditions, the resulting DAE model for the slow process dynamics has a high index.

Consider a gas phase reactor of fixed volume V_T, where a gaseous reactant A is fed at a molar flow rate F_A and temperature T_A. An exothermic irreversible reaction $A \rightarrow 2B$ yields the product B, and the product stream is withdrawn from the reactor at a flow rate F_v. The reactor pressure P is controlled using F_v as the manipulated input, through the high-gain proportional controller:

$$F_v = F_{v,nom} - K(P^* - P) \tag{2.16}$$

where $F_{v,nom}$ refers to the nominal steady-state value of F_v, and P^* is the desired value for the reactor pressure. Assuming ideal gas behavior, a dynamic model of the reactor is given by:

$$\dot{M} = F_A - F_v + k(T)My_A$$

$$\dot{y}_A = \frac{1}{M}[F_A(1 - y_A) - k(T)My_A(1 + y_A)]$$

$$\dot{T} = \frac{1}{Mc_p}[F_A c_{pA}(T_A - T) - k(T)My_A(\Delta H_r^o + (2c_{pB} - c_{pA})(T - T^o))]$$

$$0 = MRT - PV_T$$

$$0 = F_v - F_{v,nom} + K(P^* - P) \tag{2.17}$$

where M is the molar holdup, y_A is the mole fraction of A, T is the reactor temperature, c_{pA}, c_{pB} are the molar heat capacities of A and B, $c_p = c_{pA}y_A + c_{pB}(1 - y_A)$ is the molar heat capacity of the gas mixture in the reactor, and ΔH_r^o is the heat of reaction at the reference temperature T^o. The above DAE model has an index one, and is easily reduced to an ODE system by solving for F_v from the algebraic equations and substituting it in the differential equations. However, the above model exhibits a stiffness owing to the presence of the large controller gain K.

The high-gain controller essentially maintains the reactor pressure P close to the desired value P^*, after an initial fast boundary layer. Thus, a description of the slow dynamics is obtained by ignoring the fast reactor pressure dynamics and replacing the explicit controller equation with the QSS relation $P = P^*$. The resulting DAE model is:

$$\dot{M} = F_A - F_v + k(T)My_A$$

$$\dot{y}_A = \frac{1}{M}[F_A(1 - y_A) - k(T)My_A(1 + y_A)]$$

$$\dot{T} = \frac{1}{Mc_p}[F_A c_{pA}(T_A - T) - k(T)My_A(\Delta H_r^o + (2c_{pB} - c_{pA})(T - T^o))]$$

$$0 = MRT - PV_T$$

$$0 = P^* - P \tag{2.18}$$

which clearly has a high-index. In this example, it can be verified that the index of the above DAE model is two. In general, the index of a DAE model obtained by approximating such high-gain controllers with QSS conditions will be exactly one more than the relative order (degree) [64, 108] of the controlled output with respect to the manipulated input.

2.7 Flow of Incompressible Fluids

The previous examples illustrate the occurrence of high-index DAE models in chemical processes modeled as lumped parameters systems. High-index DAE systems also arise from spatial discretization of partial differential equation (PDE) models of chemical processes. For instance, the dynamic behavior of an incompressible Newtonian fluid flow in a two- or three-dimensional space is described by the standard incompressible Navier-Stokes and continuity equations:

$$\frac{\partial v}{\partial t} + v \cdot \nabla v = -\frac{\nabla p}{\rho} + \nu \nabla^2 v$$

$$\nabla \cdot v = 0 \tag{2.19}$$

where $v(t, x)$ is the velocity vector, $p(t, x)$ is the pressure and ρ, ν are the (constant) density and kinematic viscosity. A spatial discretization of the PDEs in Eq.2.19 using a finite difference or finite element method, yields the following DAE model [55]:

$$\dot{V} = M^{-1}\left[(K - N(V))V - CP - f\right]$$
$$0 = C^T V + g \qquad (2.20)$$

where M is either the identity matrix (for finite differences), or a symmetric positive definite matrix (for finite elements), and f, g depend on the boundary conditions. The discretization of the momentum balance equations yields the differential equations for the velocity vector V, while the discretization of the continuity equation yields the algebraic equations, which do not involve any of the algebraic variables $z = P$. Clearly, these algebraic equations can not be solved for pressure P, and the DAE system has a high index. More specifically, the algebraic equations in Eq.2.20 have to be differentiated once with respect to time to obtain a solution for P, and the index is two; another differentiation would yield the differential equations for P. The high index arises precisely due to the incompressibility assumption, and the analysis/simulation of incompressible Navier-Stokes equations has attracted considerable attention (see, e.g., [45, 53, 54]).

2.8 Conclusions

High-index DAE systems arise naturally in a broad class of chemical processes with fast phenomena like fast reactions, mass transfer, etc., that are approximated by the QSS conditions of phase, reaction, thermal, or pressure-equilibrium. In some cases the fast phenomena are inherent characteristics of the process, whereas in other cases they may result from high-gain controllers. Moreover, while for processes with fast mass transfer, reactions, etc., the index of the DAE models derived under QSS approximations is typically two, the index of DAE models obtained under QSS approximations for high-gain controllers depends on the corresponding relative order and could, thus, exceed two. The latter is especially significant in the modeling of process networks, where the fast dynamics of the individual units with their respective controllers are approximated with QSS conditions to obtain a DAE model for the slow dynamics of the overall network. For an example of such a reactor-separator network, see section 7.4 and [88]. The examples in this chapter were used to illustrate the different phenomena that lead to high-index DAE models. It should be mentioned that in a real process it is quite likely that several of these phenomena are present simultaneously.

A common feature of all these processes is the presence of large parameters in the detailed models, e.g., large mass transfer/reaction rate coefficients or a large controller gain, which exhibit a stiffness/time-scale multiplicity. We shall discuss this connection between DAE systems and stiff ODE systems with large parameters within the framework of singular perturbations in Chapter 6.

3. Feedback Control of Regular DAE systems

3.1 Introduction

A natural approach to address the feedback control of nonlinear DAE systems is to derive an explicit representation of the underlying ODE system (i.e., a state-space realization), and use it as the basis for the controller design. This allows utilizing the extensive machinery available for the control of nonlinear ODE systems [64,70,80,108], while systematically accounting for the differences between DAE and ODE systems. Clearly, this approach is feasible for *regular* DAE systems (see Section 1.3.4 for the definition), for which the requisite state-space realization exists independently of the controller design.

In this chapter, we address the local stabilization and output tracking through state feedback for high-index DAE systems that are regular. Initially, we present an algorithm that allows a precise characterization of the class of DAE systems that are regular and yields a state-space realization of such systems. The derived state-space realization is then used as the basis for the synthesis of a feedback controller that induces a well-characterized input/output response with stability in the closed-loop system.

3.2 Preliminaries

We consider nonlinear multi-input multi-output (MIMO) DAE systems with the semiexplicit description:

$$
\begin{aligned}
\dot{x} &= f(x) + b(x)z + g(x)u \\
0 &= k(x) + l(x)z + c(x)u \\
y_i &= h_i(x), \quad i = 1, \dots, m
\end{aligned}
\tag{3.1}
$$

where $x \in \mathcal{X} \subset \mathbb{R}^n$ is the vector of differential variables, $z \in \mathcal{Z} \subset \mathbb{R}^p$ is the vector of algebraic variables ($\mathcal{X}$, $\mathcal{Z}$ are open connected sets), $u \in \mathbb{R}^m$ is the vector of manipulated inputs and y_i is the ith output, which is a smooth function of the differential variables.

We focus on DAE systems of Eq.3.1 that have a high index, i.e., $l(x)$ is singular on $\mathcal{X}$. For such systems, the algebraic variables z are implicitly determined by the singular algebraic equations as a function of the differential variables x, the input variables u, and possibly their time-derivatives. It should be mentioned that the

31

algebraic variables z can not be manipulated in an arbitrary manner (only the input variables u can be manipulated as desired). This fact differentiates the problem of controller design for high-index DAE systems of Eq.3.1 from the control of constrained ODE systems in, for example, [22]. For, if it were possible to directly manipulate the variables z, then the control of the system in Eq.3.1 would simply be a multivariable control problem for an ODE system in the variables x with the input vector $[u^T \ z^T]^T$ and an extended set of outputs including the ones given by the algebraic equations, i.e., $\tilde{y} = k(x) + l(x)z + c(x)u$, which are always zero.

The fact that only the inputs u can be manipulated directly to stabilize the DAE system and regulate the outputs y_i, while the algebraic variables z are uniquely determined for a given u by the algebraic equations, suggests the following sequential procedure for controller design:

1. Derive a state-space realization of the DAE system, i.e., a set of differential equations for x that describe the dynamics of the system subject to the underlying algebraic constraints.

2. Formulate and solve a controller synthesis problem on the basis of the derived state-space realization using results for ODE systems.

The derivation of a state-space realization of the DAE system in Eq.3.1 involves:

(i) identifying the underlying algebraic constraints in x imposed by the singular algebraic equations; these constraints specify the state space region where the differential variables x evolve, and

(ii) obtaining a solution for the algebraic variables z in terms of x and u that is consistent with the identified algebraic constraints in x.

In this chapter, we focus on the class of DAE systems in Eq.3.1 that are regular (see Section 1.3.4 for the definition), i.e., those for which the underlying constraints in the differential variables x do not depend on the manipulated inputs u, and the the state space of the constrained system is control-invariant. An algorithm for the characterization of the class of DAE systems of Eq.3.1 that are regular, and the derivation of state-space realizations of such systems, is presented in the following section.

3.3 Derivation of State-Space Realizations

Before addressing the derivation of state-space realizations for a general DAE system of Eq.1.9, consider, for simplicity, a DAE system of Eq.1.9 with one algebraic equation and variable ($m = 1$). Clearly, this DAE system has a high index only if $l(x) \equiv 0$, i.e., the algebraic equation is as follows:

$$0 = k(x) + c(x)u \qquad (3.2)$$

For such a high-index DAE system, the algebraic equation in Eq.3.2 denotes a constraint in the differential variables x, which possibly involves the inputs u if $c(x) \neq 0$. For now, we will focus on regular systems, for which we must have $c(x) = 0$ and the algebraic equation is as follows:

$$0 = k(x) \tag{3.3}$$

The problem of deriving a state-space realization in this case can be conveniently couched as the problem of specifying the zero dynamics [64] of the dynamic system:

$$\dot{x} = f(x) + b(x)z + g(x)u \tag{3.4}$$

with the "auxiliary" output $\tilde{y} = k(x)$, viewing the algebraic variable z as an auxiliary input. Then, assuming that

(i) there exists an integer s such that $L_b L_f^i k(x) \equiv 0$, for $i < s-1$ and $L_b L_f^{s-1} k(x) \neq 0$ on $\mathcal{X}$, and

(ii) $L_g L_f^i k(x) \equiv 0$ for $i = 0, \ldots, s-2$,

where $L_b L_f^i k(x)$, $L_g L_f^i k(x)$ denote standard Lie derivatives (for a definition see [64] and the Notation section at the end of this chapter), and differentiating Eq.3.3 s times with respect to time, we obtain the following set of equations:

$$0 = L_f k(x) \tag{3.5}$$
$$0 = L_f^2 k(x) \tag{3.6}$$
$$\vdots \tag{3.7}$$
$$0 = L_f^{s-1} k(x) \tag{3.8}$$
$$0 = L_f^s k(x) + L_b L_f^{s-1} k(x)z + L_g L_f^{s-1} k(x)u \tag{3.9}$$

The algebraic equations in Eq.3.3,3.5-3.8 denote constraints in the differential variables x. It can be shown that the gradient covector fields for these constraints are linearly independent [64], and thus, they specify the following state-space region where x evolves:

$$\mathcal{M} = \left\{ x \in \mathcal{X} : \begin{array}{c} k(x) = 0 \\ \vdots \\ L_f^{s-1} k(x) = 0 \end{array} \right\} \tag{3.10}$$

which is a smooth manifold of dimension $n - s$. Moreover, the final algebraic equation (Eq.3.9) uniquely determines the solution for z:

$$z = \frac{-1}{L_b L_f^{s-1} k(x)} \left(L_f^s k(x) + L_g L_f^{s-1} k(x)u \right) \tag{3.11}$$

Substituting this solution for z in the differential equations for x in Eq.1.9, yields a state-space realization:

$$\dot{x} = \left(f(x) - \frac{L_f^s k(x)}{L_b L_f^{s-1} k(x)} b(x) \right) + \left(g(x) - \frac{L_g L_f^{s-1} k(x)}{L_b L_f^{s-1} k(x)} b(x) \right) u \qquad (3.12)$$
$$y_i = h_i(x), \quad i = 1, \ldots, m$$

of the DAE system on the constrained state space $\mathcal{M}$.

In the above derivation of the state-space realization, condition (i) essentially ensures that a well-defined finite index exists, while condition (ii) ensures that the underlying algebraic constraints in x (Eq.3.3-3.8) are independent of the manipulated inputs u, i.e., the DAE system is regular. Under analogous conditions for a class of DAE systems with multiple algebraic equations and variables, the derivation of a state-space realization of the DAE system was addressed in [75, 79, 102] through repetitive differentiations of the algebraic equations until the resulting equations could be solved for z.

In what follows, we address the problem of deriving state-space realizations of a general DAE system in Eq.3.1 with multiple algebraic equations and variables. Similar to the case of one algebraic equation, the problem can be viewed as the one of specifying the zero dynamics of the dynamic system in Eq.3.4, viewing the algebraic equations

$$k(x) + l(x)z + c(x)u = 0 \qquad (3.13)$$

as a set of auxiliary outputs $\widetilde{y}$ and the algebraic variables z as the corresponding set of auxiliary inputs. Note, however, that in this general setting, the "outputs" $\widetilde{y}$ depend explicitly on the "inputs" z, and this input/output relation is singular. In the next section, an algorithm is presented that addresses the derivation of state-space realizations of a general regular DAE system of Eq.1.9, through the specification of the aforementioned zero dynamics (see also [87] for a variant of this algorithm). It is based on Hirschorn's inversion algorithm [61], which was introduced in the context of calculating the inverse of a nonlinear MIMO ODE system with a singular input/output map (in the sense of singularity of the characteristic/decoupling matrix). In brief, the inversion algorithm involves successive elementary row operations on the outputs of the ODE system to localize the singularity in specific outputs, followed by a differentiation of these outputs, until a nonsingular input/output relation is obtained that can be solved for the inputs. Similarly, the proposed algorithm involves a sequence of algebraic operations on the singular algebraic equations to identify the underlying constraints in x, and a differentiation of these constraints, to obtain a solution for z.

3.3.1 Algorithm for reconstruction of algebraic variables

Iteration 1:

Consider the algebraic equations in Eq.3.13, where rank $l(x) = p_1 < p$ on $\mathcal{X}$ (this region can be redefined appropriately to exclude points of singularity for the above as well as the subsequent rank conditions). Then, there exists a smooth nonsingular $p \times p$ matrix $E^1(x)$, which

(i) rearranges the rows of the matrix $l(x)$ such that the first p_1 rows of $E^1(x)l(x)$ are linearly independent, and

(ii) reduces the last $p - p_1$ rows of $E^1(x)l(x)$ to zero.

Step 1. Premultiply the algebraic equations in Eq.3.13 by such a matrix $E^1(x)$, to obtain the following set of equations:

$$0 = \begin{bmatrix} \overline{k}^1(x) \\ \mathbf{k}^1(x) \end{bmatrix} + \begin{bmatrix} \overline{l}^1(x) \\ 0 \end{bmatrix} z + \begin{bmatrix} \overline{c}^1(x) \\ \hat{c}^1(x) \end{bmatrix} u \tag{3.14}$$

where $\overline{l}^1(x)$ is a $p_1 \times p$ matrix with full row rank, $\overline{k}^1(x)$ and $\mathbf{k}^1(x)$ are vector fields of dimensions p_1 and $(p - p_1)$, and $\overline{c}^1(x)$, $\hat{c}^1(x)$ are matrices of dimensions $p_1 \times m$ and $(p - p_1) \times m$, respectively. Note that the last $(p - p_1)$ algebraic equations in Eq.3.14, $0 = \mathbf{k}^1(x) + \hat{c}^1(x)u$, denote constraints in x. For the DAE system of Eq.3.1 to be regular, these constraints must be independent of u. Clearly, this is true if only if $\hat{c}^1(x) \equiv 0$, or equivalently,

$$\text{rank } [l(x) \ c(x)] = \text{rank } l(x) \tag{3.15}$$

Step 2. Differentiate the last $p - p_1$ equations of Eq.3.14, i.e., the constraints $0 = \mathbf{k}^1(x)$, once, to obtain the following new set of algebraic equations:

$$0 = \begin{bmatrix} \overline{k}^1(x) \\ \widetilde{k}^2(x) \end{bmatrix} + \begin{bmatrix} \overline{l}^1(x) \\ \widetilde{l}^2(x) \end{bmatrix} z + \begin{bmatrix} \overline{c}^1(x) \\ \widetilde{c}^2(x) \end{bmatrix} u \tag{3.16}$$

where

$$\widetilde{k}^2(x) = [L_f \mathbf{k}_1^1(x) \ \cdots \ L_f \mathbf{k}_{p-p_1}^1(x)]^T$$

and

$$\widetilde{l}^2(x) = \begin{bmatrix} L_{b_1} \mathbf{k}_1^1(x) & \cdots & L_{b_p} \mathbf{k}_1^1(x) \\ \vdots & & \vdots \\ L_{b_1} \mathbf{k}_{p-p_1}^1(x) & \cdots & L_{b_p} \mathbf{k}_{p-p_1}^1(x) \end{bmatrix}, \quad \widetilde{c}^2(x) = \begin{bmatrix} L_{g_1} \mathbf{k}_1^1(x) & \cdots & L_{g_m} \mathbf{k}_1^1(x) \\ \vdots & & \vdots \\ L_{g_1} \mathbf{k}_{p-p_1}^1(x) & \cdots & L_{g_m} \mathbf{k}_{p-p_1}^1(x) \end{bmatrix}$$

In the above relations, $\mathbf{k}_i^1(x)$ denotes the ith component of the vector field $\mathbf{k}^1(x)$, $b_j(x), g_j(x)$ denote the jth column vectors of the corresponding matrices, and $L_f \mathbf{k}_i^1(x)$, $L_{b_j} \mathbf{k}_i^1(x)$, $L_{g_j} \mathbf{k}_i^1(x)$ denote standard Lie derivatives.

Step 3. Evaluate the rank p_2 of the matrix:

$$\begin{bmatrix} \overline{l}^1(x) \\ \widetilde{l}^2(x) \end{bmatrix} \tag{3.17}$$

If $p_2 = p$ then stop. If $p_2 < p$, then proceed to the next iteration, starting with the set of algebraic equations in Eq.3.16.

Iteration q $(q \geq 2)$:

Consider the following set of algebraic equations obtained from iteration $q - 1$:

$$0 = \begin{bmatrix} \bar{k}^{q-1}(x) \\ \tilde{k}^q(x) \end{bmatrix} + \begin{bmatrix} \bar{l}^{q-1}(x) \\ \tilde{l}^q(x) \end{bmatrix} z + \begin{bmatrix} \bar{c}^{q-1}(x) \\ \tilde{c}^q(x) \end{bmatrix} u \qquad (3.18)$$

with

$$\text{rank} \begin{bmatrix} \bar{l}^{q-1}(x) \\ \tilde{l}^q(x) \end{bmatrix} = p_q < p$$

Then, there exists a $p \times p$ permutation matrix $E^{q,1}$ which rearranges the rows of $\tilde{l}^q(x)$ such that the first p_q rows of

$$E^{q,1} \begin{bmatrix} \bar{l}^{q-1} \\ \tilde{l}^q(x) \end{bmatrix}$$

are linearly independent. Furthermore, there exists a $p \times p$ nonsingular matrix $E^{q,2}(x)$ of the form:

$$E^{q,2}(x) = \begin{bmatrix} I_{p_q} & 0 \\ R^q(x) & S^q(x) \end{bmatrix}$$

such that

$$E^{q,2}(x)E^{q,1} \begin{bmatrix} \bar{l}^{q-1}(x) \\ \tilde{l}^q(x) \end{bmatrix} = \begin{bmatrix} \bar{l}^q(x) \\ 0 \end{bmatrix}$$

where the $p_q \times p$ matrix $\bar{l}^q(x)$ has full row rank.

Step 1. Premultiply Eq.3.18 by the matrix $E^q(x) = E^{q,2}(x)E^{q,1}$ to obtain:

$$0 = \begin{bmatrix} \bar{k}^q(x) \\ \mathbf{k}^q(x) \end{bmatrix} + \begin{bmatrix} \bar{l}^q(x) \\ 0 \end{bmatrix} z + \begin{bmatrix} \bar{c}^q(x) \\ \hat{c}^q(x) \end{bmatrix} u \qquad (3.19)$$

where $\bar{c}^q(x)$ and $\hat{c}^q(x)$ are matrices of dimensions $p_q \times m$ and $(p - p_q) \times m$, and $\bar{k}^q, \mathbf{k}^q(x)$ are vector fields of dimensions p_q and $(p - p_q)$, respectively. Clearly, for the DAE system to be regular, the constraints in x, $0 = \mathbf{k}^q(x) + \hat{c}^q(x)u$, must be independent of the inputs u, i.e., $\hat{c}^q(x) \equiv 0$, or equivalently,

$$\text{rank} \begin{bmatrix} \bar{l}^{q-1}(x) & \bar{c}^{q-1}(x) \\ \tilde{l}^q(x) & \tilde{c}^q(x) \end{bmatrix} = \text{rank} \begin{bmatrix} \bar{l}^{q-1}(x) \\ \tilde{l}^q(x) \end{bmatrix} \qquad (3.20)$$

Step 2. Differentiate the last $p - p_q$ equations of Eq.3.19, i.e., the constraints $0 = \mathbf{k}^q(x)$, once, to obtain the following set of algebraic equations:

$$0 = \begin{bmatrix} \overline{k}^q(x) \\ \widetilde{k}^{q+1}(x) \end{bmatrix} + \begin{bmatrix} \overline{l}^q(x) \\ \widetilde{l}^{q+1}(x) \end{bmatrix} z + \begin{bmatrix} \overline{c}^q(x) \\ \widetilde{c}^{q+1}(x) \end{bmatrix} u \tag{3.21}$$

Step 3. Evaluate the rank p_{q+1} of the matrix:

$$\begin{bmatrix} \overline{l}^q(x) \\ \widetilde{l}^{q+1}(x) \end{bmatrix}$$

If $p_{q+1} = p$ then stop, else repeat the above steps for the next iteration, starting with the set of algebraic equations (Eq.3.21).

By construction, the algorithm generates a sequence of integers $p_1 \leq p_2 \leq \cdots \leq p$ (it is assumed that in each iteration i, the rank p_i is constant on $\mathcal{X}$). For a DAE system with a finite index ν_d, the algorithm converges after a finite number of iterations s, with the following final set of algebraic equations:

$$0 = \begin{bmatrix} \overline{k}^s(x) \\ \widetilde{k}^{s+1}(x) \end{bmatrix} + \begin{bmatrix} \overline{l}^s(x) \\ \widetilde{l}^{s+1}(x) \end{bmatrix} z + \begin{bmatrix} \overline{c}^s(x) \\ \widetilde{c}^{s+1}(x) \end{bmatrix} u \tag{3.22}$$

where the $p \times p$ matrix.

$$\begin{bmatrix} \overline{l}^s(x) \\ \widetilde{l}^{s+1}(x) \end{bmatrix}$$

has full rank, i.e., $p_{s+1} = p$. The above final set of algebraic equations can be solved for the algebraic variables z as a function of the differential variables x and the manipulated inputs u, to obtain:

$$z = - \begin{bmatrix} \overline{l}^s(x) \\ \widetilde{l}^{s+1}(x) \end{bmatrix}^{-1} \left\{ \begin{bmatrix} \overline{k}^s(x) \\ \widetilde{k}^{s+1}(x) \end{bmatrix} + \begin{bmatrix} \overline{c}^s(x) \\ \widetilde{c}^{s+1}(x) \end{bmatrix} u \right\} \tag{3.23}$$

Note that for the DAE system in Eq.3.1, altogether s differentiations were required to obtain a solution for the algebraic variables z. Differentiating the solution for z in Eq.3.23 once more, a set of differential equations for z can be obtained. Hence, the index ν_d of the DAE system is exactly $s + 1$.

The algorithm also identifies a set of $\sum_{i=1}^{s}(p - p_i)$ algebraic constraints in the differential variables x:

$$\mathbf{k}(x) = \begin{bmatrix} \mathbf{k}^1(x) \\ \vdots \\ \mathbf{k}^s(x) \end{bmatrix} = 0 \tag{3.24}$$

Proposition 3.1 that follows, establishes the linear independence of these constraints.

Proposition 3.1: *Consider the DAE system of Eq.3.1 for which the proposed algorithm converges after s iterations. Then, the $\sum_{i=1}^{s}(p - p_i)$ constraints $\mathbf{k}(x) = 0$ in Eq.3.24 are linearly independent, in the sense that the gradient covector fields:*

$$d\,\mathbf{k}_i(x) = \left[\frac{\partial \mathbf{k}_i(x)}{\partial x_1} \quad \cdots \quad \frac{\partial \mathbf{k}_i(x)}{\partial x_n} \right], \quad i = 1, \ldots, \sum_{j=1}^{s}(p - p_j)$$

are linearly independent.

Proof: Consider a DAE system (Eq.3.1) for which the algorithm converges after s iterations, i.e., $p_{s+1} = p$. The aim is to prove that the $\sum_{i=1}^{s}(p - p_i)$ constraints $\mathbf{k}_j^i(x) = 0$, $i = 1, \ldots, s$; $j = 1, \ldots, (p - p_i)$ are linearly independent, in the sense that the gradient covector fields $d\mathbf{k}_j^i(x)$ are linearly independent. A proof of this is given following an inductive procedure similar to that of [130]. The key idea of the proof is to show that if, in any iteration q, a covector field $d\mathbf{k}_r^q(x)$ is linearly dependent on the covector fields $d\mathbf{k}_1^1(x), \ldots, d\mathbf{k}_{p-p_1}^1(x), \ldots, d\mathbf{k}_1^q(x), \ldots, d\mathbf{k}_{r-1}^q(x), d\mathbf{k}_{r+1}^q(x)$, $\ldots, d\mathbf{k}_{p-p_q}^q(x)$, then

(*i*) $p_{q+1} < p$, and

(*ii*) in iteration $q + 1$, there exists an $r \in [1, (p - p_{q+1})]$ such that $d\mathbf{k}_r^{q+1}(x)$ is linearly dependent on $d\mathbf{k}_1^1(x), \ldots, d\mathbf{k}_{p-p_1}^1(x), \ldots, d\mathbf{k}_1^{q+1}(x), \ldots, d\mathbf{k}_{r-1}^{q+1}(x), d\mathbf{k}_{r+1}^{q+1}(x)$, $\ldots, d\mathbf{k}_{p-p_{q+1}}^{q+1}(x)$,

thereby implying by induction that $p_{s+1} < p$, which clearly is a contradiction.

Let iteration q be the first iteration when the constraints $\mathbf{k}^1(x) = 0, \ldots, \mathbf{k}^q(x) = 0$ are not linearly independent. More specifically, for some $r \in [1, (p - p_q)]$, the gradient covector field $d\mathbf{k}_r^q(x)$ is linearly dependent on the gradients of the remaining scalar functions, in the set:

$$\mathcal{K}^q = \left\{ \begin{array}{c} \mathbf{k}_1^1(x), \ldots, \mathbf{k}_{p-p_1}^1(x), \\ \mathbf{k}_1^2(x), \ldots, \mathbf{k}_{p-p_2}^2(x), \\ \vdots \\ \mathbf{k}_1^q(x), \ldots, \mathbf{k}_{r-1}^q(x), \mathbf{k}_{r+1}^q(x), \ldots, \mathbf{k}_{p-p_q}^q(x) \end{array} \right\}$$

Define the following two disjoint subsets of $\mathcal{K}^q$:

$\mathcal{K}^{q,I} = \{\, \mathbf{k}_j^i(x) \in \mathcal{K}^q \; : \; d\mathbf{k}_j^i(x)b(x)$ is a row of the matrix $\bar{l}^{q+1}(x) \,\}$

$\mathcal{K}^{q,II} = \{\, \mathbf{k}_j^i(x) \in \mathcal{K}^q \; : \; d\mathbf{k}_j^i(x)b(x)$ is a linear combination of the rows of $\bar{l}^{q+1}(x) \,\}$

and the corresponding vector fields $\mathbf{K}^{q,I}(x)$, $\mathbf{K}^{q,II}(x)$ whose components are the scalar elements of the sets $\mathcal{K}^{q,I}$ and $\mathcal{K}^{q,II}$, respectively. Then, by definition:

$$\bar{l}^{q+1}(x) = \left[\begin{array}{c} \bar{l}^1(x) \\ d\mathbf{K}^{q,I}(x)b(x) \end{array} \right],$$

and owing to the linear dependence of $d\mathbf{k}_r^q(x)$ on the gradients of the scalar functions $\mathbf{k}_j^i$ in the set $\mathcal{K}^q$, there exist row vectors $R_1(x)$ and $R_2(x)$ such that:

$$d\mathbf{k}_r^q(x) = R_1(x)d\mathbf{K}^{q,I}(x) + R_2(x)d\mathbf{K}^{q,II}(x) \tag{3.25}$$

Moreover, by the definition of the two subsets $\mathcal{K}^{q,I}, \mathcal{K}^{q,II}$, and the rank condition in Eq.3.20, there exists a matrix $S_1(x)$ such that:

$$dK^{q,II}(x)b(x) = S_1(x)\,\tilde{l}^{q+1}(x)$$
$$dK^{q,II}(x)g(x) = S_1(x)\,\tilde{c}^{q+1}(x) = S_1(x)\begin{bmatrix} \tilde{c}^1(x) \\ dK^{q,I}(x)g(x) \end{bmatrix} \tag{3.26}$$

From Eq.3.25 and Eq.3.26, it follows that:

$$\tilde{l}_r^{q+1}(x) = d\mathbf{k}_r^q(x)b(x) = R_1(x)dK^{q,I}(x)b(x) + R_2(x)S_1(x)\tilde{l}^{q+1}(x) \tag{3.27}$$

proving, thus, that the row vector $\tilde{l}_r^{q+1}(x)$ is linearly dependent on the rows of the matrix $\tilde{l}^{q+1}(x)$, i.e., $p_{q+1} < p$.

The relation in Eq.3.27 implies that in step 1 of iteration $q + 1$, the row vector $\tilde{l}_r^{q+1}(x)$ (and the row vector $\tilde{c}_r^{q+1}(x) = d\mathbf{k}_r^q(x)g(x)$) can be reduced to zero through elementary row operations, to obtain the corresponding algebraic constraint in x:

$$0 = \mathbf{k}_r^{q+1}(x) = \left(d\mathbf{k}_r^q(x)f(x) - R_1(x)dK^{q,I}(x)f(x)\right) - R_2(x)S_1(x)\begin{bmatrix} \bar{k}^1(x) \\ dK^{q,I}(x)f(x) \end{bmatrix}$$
$$= R_2(x)dK^{q,II}(x)f(x) - R_2(x)S_1(x)\begin{bmatrix} \bar{k}^1(x) \\ dK^{q,I}(x)f(x) \end{bmatrix} \tag{3.28}$$

Also, by definition, the $p_{q+1} \times p$ matrix $\bar{l}^{q+1}(x)$ has full row rank, which implies that there exists a matrix:

$$\bar{l}^{q+1\dagger}(x) = \bar{l}^{q+1^T}(x)\left[\bar{l}^{q+1}(x)\bar{l}^{q+1^T}(x)\right]^{-1}$$

such that $\bar{l}^{q+1}(x)\bar{l}^{q+1\dagger}(x) = I_{p_{q+1}}$. Thus, from Eq.3.26 it follows that:

$$R_2(x)S_1(x) = R_2(x)dK^{q,II}(x)b(x)\bar{l}^{q+1\dagger}(x)$$

Substituting the above relation for $R_2(x)S_1(x)$ in Eq.3.28 the following relation is obtained:

$$\mathbf{k}_r^{q+1}(x) = R_2(x)dK^{q,II}(x)\left\{f(x) - b(x)\bar{l}^{q+1\dagger}(x)\begin{bmatrix} \bar{k}^1(x) \\ dK^{q,I}(x)f(x) \end{bmatrix}\right\} \tag{3.29}$$

Furthermore, consider the vector field $\mathbf{k}^{q,II}(x)$ of dimension $p - p_{q+1} - 1$, comprised of the elements of $\mathbf{k}^q(x)$ that belong to the set $\mathcal{K}^{q,II}$. Then, by the definition of the set $\mathcal{K}^{q,II}$, the rows of the matrix matrix:

$$d\mathbf{k}^{q,II}(x)b(x) = \begin{bmatrix} d\mathbf{k}_1^{q,II}(x)b(x) \\ \vdots \\ d\mathbf{k}_{r-1}^{q,II}(x)b(x) \\ d\mathbf{k}_{r+1}^{q,II}(x)b(x) \\ \vdots \\ d\mathbf{k}_{p-p_{q+1}}^{q,II}(x)b(x) \end{bmatrix}$$

are linearly dependent on the rows of $\bar{l}^{q+1}(x)$ and, thus, there exists a matrix $\gamma^q(x)$ such that:

$$d\mathbf{k}^{q,II}(x)b(x) = \gamma^q(x)\bar{l}^{q+1}(x)$$

Clearly,

$$\gamma^q(x) = d\mathbf{k}^{q,II}(x)b(x)\bar{l}^{q+1\dagger}(x)$$

Thus, the rows of $d\mathbf{k}^{q,II}(x)b(x)$ (and correspondingly the rows of $d\mathbf{k}^{q,II}(x)g(x)$) can be reduced to zero through row operations in step 1 of iteration $q + 1$, yielding the remaining $p - p_{q+1} - 1$ algebraic constraints in x (besides the constraint in Eq.3.28):

$$0 = \begin{bmatrix} \mathbf{k}_1^{q+1}(x) \\ \vdots \\ \mathbf{k}_{r-1}^{q+1}(x) \\ \mathbf{k}_{r-1}^{q+1}(x) \\ \vdots \\ \mathbf{k}_{p-p_{q+1}}^{q+1}(x) \end{bmatrix} = d\mathbf{k}^{q,II}(x)f(x) - \gamma^q(x)\begin{bmatrix} \bar{k}^1(x) \\ dK^{q,I}(x)f(x) \end{bmatrix}$$

$$\hspace{2cm} (3.30)$$

$$= d\mathbf{k}^{q,II}(x)\left\{ f(x) - b(x)\bar{l}^{q+1\dagger}(x)\begin{bmatrix} \bar{k}^1(x) \\ dK^{q,I}(x)f(x) \end{bmatrix} \right\}$$

From the relations in Eq.3.29 and Eq.3.30, and the definition of $\mathbf{K}^{q,II}(x)$ it follows that:

$$\mathbf{k}_r^{q+1}(x) = R_2(x)\alpha^{q+1}(x)$$
$$= \sum_{i=1}^{\lambda} R_{2,i}(x)\alpha_i^{q+1}(x) \hspace{1cm} (3.31)$$

where $\alpha^{q+1}(x) = [\alpha_1^{q+1}(x) \cdots \alpha_\lambda^{q+1}(x)]^T$ such that:

$$\alpha_i^{q+1}(x) \in \mathcal{K}^{q+1}$$

with the set $\mathcal{K}^{q+1}$ defined in an analogous fashion to $\mathcal{K}^q$ ($\mathcal{K}^q \subseteq \mathcal{K}^{q+1}$), and $R_{2,i}(x)$ is the ith component of the row vector $R_2(x)$.

Claim: The gradient covector fields $dR_{2,i}(x)$ are linearly dependent on the row vectors of $d\mathbf{K}^{q+1,I}(x)$ and $d\mathbf{K}^{q+1,II}(x)$.

If the above claim is true, then it is straightforward to verify that the gradient covector field $d\mathbf{k}_r^{q+1}(x)$ is linearly dependent on the row vectors of $d\mathbf{K}^{q+1,I}(x), d\mathbf{K}^{q+1,II}(x)$, i.e., in iteration $q+1$, the constraint $\mathbf{k}_r^{q+1}(x) = 0$ is linearly dependent on the remaining constraints corresponding to the scalar functions in the set $\mathcal{K}^{q+1}$. Thus, the whole argument can be repeated for iteration $(q+1)$ implying that $p_{q+2} < p$ and by induction, it follows that $p_{s+1} < p$, thereby leading to a contradiction. $\qquad\square$

Proof of claim: Consider the relation in Eq.3.25. Without loss of generality, it is assumed that the gradient covector fields corresponding to the components of the vector fields $\mathbf{K}^{q,I}(x)$ and $\mathbf{K}^{q,II}(x)$ are linearly independent. Note that this is definitely true for the vector field $\mathbf{K}^{q,I}(x)$, owing to its definition. If there are some components of $\mathbf{K}^{q,II}(x)$, for which the gradient covector fields are linearly dependent on the others, then the corresponding component of $R_2(x)$ will be identically equal to zero. Let $\mu = \mu_1 + \mu_2$, where μ_1, μ_2 denote the dimensions of the vector fields $\mathbf{K}^{q,I}(x)$ and $\mathbf{K}^{q,II}(x)$, respectively. Then it is always possible to find $n - \mu$ smooth scalar functions $\psi_1(x), \ldots, \psi_{n-\mu}(x)$ such that:

$$
\zeta = \begin{bmatrix} \zeta_1 \\ \vdots \\ \zeta_{\mu_1} \\ \zeta_{\mu_1+1} \\ \vdots \\ \zeta_\mu \\ \zeta_{\mu+1} \\ \vdots \\ \zeta_n \end{bmatrix} = \begin{bmatrix} \mathbf{K}_1^{q,I}(x) \\ \vdots \\ \mathbf{K}_{\mu_1}^{q,I} \\ \mathbf{K}_1^{q,II}(x) \\ \vdots \\ \mathbf{K}_{\mu_2}^{q,II}(x) \\ \psi_1(x) \\ \vdots \\ \psi_{n-\mu}(x) \end{bmatrix}
$$

is a valid local diffeomorphism ($\mathbf{K}_i^{q,I}(x), \mathbf{K}_i^{q,II}(x)$ denote the ith components of the corresponding vector fields). In the new coordinates ζ, the gradient covector fields $d\mathbf{K}_i^{q,I}(\zeta)$ take the form $[0 \cdots 0\ 1\ 0 \cdots 0]$ with the nonzero entry "1" in the ith position, while the gradient covector fields $d\mathbf{K}_i^{q,II}(\zeta)$ take the form $[0 \cdots 0\ 1\ 0 \cdots 0]$ with the nonzero entry "1" in the $(\mu_1 + i)$th position. Clearly, the covector field $d\mathbf{k}_r^q(\zeta)$ must have the form $[* \cdots *\ 0 \cdots 0]$, i.e., only the first μ entries can be nonzero (denoted by "$*$"). Thus, it follows that $\left(\dfrac{\partial R_{2,j}(\zeta)}{\partial \zeta_l}\right) \equiv 0$, $j = 1, \ldots, \mu_2$; $l = (\mu + 1), \ldots, n$ which implies that the gradient covector fields $dR_{2,j}(\zeta)$ must be linearly dependent on the rows of $d\mathbf{K}^{q,I}(\zeta)$ and $d\mathbf{K}^{q,II}(\zeta)$, and consequently, on the rows of $d\mathbf{K}^{q+1,I}(\zeta)$ and $d\mathbf{K}^{q+1,II}(\zeta)$. This completes the proof of the claim. $\qquad\square$

Proposition 3.1 leads to the following observations for the DAE system of Eq.3.1:

1. Given $x \in \mathbb{R}^n$, the linear independence of the constraints in Eq.3.24 implies that $\sum_{i=1}^{s} (p - p_i) \leq n$.

2. The algebraic constraints in Eq.3.24 completely specify the state space $\mathcal{M} \subset \mathbb{R}^n$ where the differential variables x of the constrained DAE system (Eq.3.1) evolve. More specifically,

$$\mathcal{M} = \{x \in \mathcal{X} \subset \mathbb{R}^n \ : \ \mathbf{k}(x) = 0\} \tag{3.32}$$

which, given the linear independence of $d\mathbf{k}_i(x)$, $i = 1, \ldots, \sum_{i=1}^{s}(p - p_i)$, is a smooth manifold of dimension $\kappa = n - \sum_{i=1}^{s}(p - p_i)$.

3. The scalar functions $\mathbf{k}_i(x)$, $i = 1, \ldots, \sum_{i=1}^{s}(p - p_i)$ can be used as a part of a local coordinate change to derive a state-space realization of the DAE system in Eq.3.1, such that the constraints in Eq.3.24 are trivially satisfied. Details of the coordinate change and the resulting state-space realization are given in the next subsection.

From the algorithm it is also clear that the rank conditions in Eq.3.15 and Eq.3.20 characterize the class of DAE systems in Eq.3.1 for which the underlying constraints in x are independent of u, i.e., systems that are regular. This result is formally stated in the following lemma.

Lemma 3.1: *Consider a DAE system of Eq.3.1 for which the algorithm converges in s iterations and the index is $\nu_d = s + 1$. Then, the DAE system is regular, if and only if, the conditions in Eq.3.15 and Eq.3.20 are satisfied in iteration 1 and iterations $q \geq 2$, respectively.*

Proof: Consider the DAE system of Eq.3.1, for which the algorithm converges in $s = \nu_d - 1$ iterations. From the algorithm, it is clear that for a specific choice of the matrices $E^i(x)$, $i = 1, \ldots, s$ (it is not unique), the $(p - p_i)$ constraints in x identified in each iteration i of the algorithm, are independent of the manipulated inputs u, if and only if the conditions in Eq.3.15 and Eq.3.20 are satisfied. Following an approach similar to [61], it can be shown that the integer ranks $p_1, \ldots, p_{s+1}$ are the same, and the rank conditions in Eq.3.15 and Eq.3.20 are satisfied, irrespectively of the choice of $E^i(x)$. Thus, the conditions in Eq.3.15, 3.20 completely specify the class of high-index DAE systems of Eq.3.1 for which the underlying algebraic constraints in x are independent of u, and the state space $\mathcal{M}$ (Eq.3.32) is control-invariant. This completes the proof. □

Remark 3.1: Consider the DAE system of Eq.3.1, for which the algorithm converged after s iterations and the index is $\nu_d = s + 1$. Then, according to [61], the integer s denotes the relative order of the auxiliary outputs $\tilde{y}$ with respect to the auxiliary inputs z for the system of Eq.3.4. This observation establishes a transparent relation between the concept of relative order and the concept of index.

3.3.2 State-space realizations of DAE system

The specification of the constrained state space $\mathcal{M}$ (Eq.3.32) and the solution for the algebraic variables z (Eq.3.23) allow the derivation of a state-space realization of the DAE system in Eq.3.1. The resulting state-space realization is given the following proposition.

Proposition 3.2: *Consider the DAE system of Eq.3.1 for which the proposed algorithm converges after s iterations, with the constraints in Eq.3.24 and the solution for z in Eq.3.23. Then, the dynamic system:*

$$\dot{x} = \overline{f}(x) + \overline{g}(x)u$$
$$y_i = h_i(x), \quad i = 1, \ldots, m \tag{3.33}$$

where $x \in \mathcal{M} = \{x \in \mathcal{X} \ : \ \mathbf{k}(x) = 0\}$ and

$$\overline{f}(x) = f(x) - b(x) \begin{bmatrix} \overline{l}^s(x) \\ \widetilde{l}^{s+1}(x) \end{bmatrix}^{-1} \begin{bmatrix} \overline{k}^s(x) \\ \widetilde{k}^{s+1}(x) \end{bmatrix}$$

$$\overline{g}(x) = g(x) - b(x) \begin{bmatrix} \overline{l}^s(x) \\ \widetilde{l}^{s+1}(x) \end{bmatrix}^{-1} \begin{bmatrix} \overline{c}^s(x) \\ \widetilde{c}^{s+1}(x) \end{bmatrix} \tag{3.34}$$

is a state-space realization of the DAE system.

Proof: The algebraic constraints of Eq.3.22 identified by the proposed algorithm, specify the constrained manifold $\mathcal{M}$ (Eq.3.32). It is straightforward to verify that for a set of initial conditions $x(0) \in \mathcal{M}$ such that $\mathbf{k}_j^i(x(0)) = 0$, $i = 1, \ldots, s$; $j = 1, \ldots, (p - p_i)$, and z satisfying the relation in Eq.3.23, the algebraic equations in Eq.3.13 and the algebraic constraints of Eq.3.24 are satisfied for all times and for any continuous inputs $u(t)$. Thus, the differential variables x evolve in the constrained state space $\mathcal{M}$ irrespectively of any control law for u, and a direct substitution of the solution for z (Eq.3.23) in the differential equations of Eq.3.4 yields the state-space realization in Eq.3.33. □

In view of the fact that the differential variables x are constrained to evolve on the manifold $\mathcal{M}$ of dimension $\kappa = n - \sum_{i=1}^{s}(p - p_i)$, the state-space realization in Eq.3.33 is not of minimal order. Such a realization can only be obtained in appropriate transformed coordinates. More specifically, given the vector fields $\mathbf{k}^i(x)$, $i = 1, \ldots, s$ with the components $\mathbf{k}_j^i(x)$, $j = 1, \ldots, (p - p_i)$, one can always find smooth functions

$\phi_1(x), \ldots, \phi_\kappa(x)$ such that

$$
\zeta = \begin{bmatrix} \zeta_1^{(0)} \\ \vdots \\ \zeta_\kappa^{(0)} \\ \zeta^{(1)} \\ \vdots \\ \zeta^{(s)} \end{bmatrix} = T(x) = \begin{bmatrix} \phi_1(x) \\ \vdots \\ \phi_\kappa(x) \\ \mathbf{k}^1(x) \\ \vdots \\ \mathbf{k}^s(x) \end{bmatrix} \tag{3.35}
$$

is a valid local diffeomorphism. In these transformed coordinates ζ, the constraints $\mathbf{k}(x) = 0$ reduce to $\zeta^{(i)} = 0$, $i = 1, \ldots, s$, which allows obtaining the minimal-order state-space realization given in Proposition 3.3.

Proposition 3.3: *Consider the DAE system of Eq.3.1 for which the proposed algorithm converges after s iterations, with the constraints in Eq.3.24 and the solution for z in Eq.3.23. Then, the dynamic system:*

$$
\begin{aligned}
\dot{\zeta}^{(0)} &= f^{(0)}(\zeta^{(0)}) + g^{(0)}(\zeta^{(0)})u \\
y_i &= h_i(x)\big|_{x=T^{-1}(\zeta^{(0)},0,\ldots,0)}, \quad i = 1, \ldots, m
\end{aligned} \tag{3.36}
$$

is a state-space realization of the DAE system, of dimension $\kappa = (n - \sum_{i=1}^{s}(p - p_i))$, where

$$
\begin{aligned}
f^{(0)}(\zeta^{(0)}) &= \begin{bmatrix} L_{\overline{f}}\phi_1(x) \\ \vdots \\ L_{\overline{f}}\phi_\kappa(x) \end{bmatrix}_{x=T^{-1}(\zeta^{(0)},0,\ldots,0)} \\
g^{(0)}(\zeta^{(0)}) &= \begin{bmatrix} L_{\overline{g}_1}\phi_1(x) & \cdots & L_{\overline{g}_m}\phi_1(x) \\ \vdots & & \vdots \\ L_{\overline{g}_1}\phi_\kappa(x) & \cdots & L_{\overline{g}_m}\phi_\kappa(x) \end{bmatrix}_{x=T^{-1}(\zeta^{(0)},0,\ldots,0)}
\end{aligned} \tag{3.37}
$$

and $\overline{f}(x), \overline{g}(x)$ are defined in Eq.3.34.

Proof: In the transformed coordinates ζ (Eq.3.35), the differential equations in the DAE system of Eq.3.1 take the following form:

$$
\begin{aligned}
\dot{\zeta}^{(0)} &= L_f\phi(x) + L_b\phi(x)z + L_g\phi(x)u \\
\dot{\zeta}^{(1)} &= L_f\mathbf{k}^1(x) + L_b\mathbf{k}^1(x)z + L_g\mathbf{k}^1(x)u \\
&\vdots \\
\dot{\zeta}^{(s)} &= L_f\mathbf{k}^s(x) + L_b\mathbf{k}^s(x)z + L_g\mathbf{k}^s(x)u
\end{aligned} \tag{3.38}
$$

where $x = T^{-1}(\zeta^{(0)}, \zeta^{(1)}, \ldots, \zeta^{(s)})$, $L_f\phi(x) = [L_f\phi_1(x) \quad \cdots \quad L_f\phi_\kappa(x)]^T$, $L_f\mathbf{k}^q(x) = [L_f\mathbf{k}_1^q(x) \quad \cdots \quad L_f\mathbf{k}_{p-p_q}^q(x)]^T$, and $L_b\phi(x), L_g\phi(x), L_b\mathbf{k}^q(x), L_g\mathbf{k}^q(x)$ are matrices whose

44

(i, j)th components are given by $L_{b_j}\phi_i(x)$, $L_{g_j}\phi_i(x)$, $L_{b_j}\mathbf{k}_i^q(x)$ and $L_{g_j}\mathbf{k}_i^q(x)$, respectively. Substituting the solution for z (Eq.3.23) in the above differential equations, the following set of differential equations is obtained:

$$\dot{\zeta}^{(0)} = L_{\overline{f}}\phi(x) + L_{\overline{g}}\phi(x)u$$
$$\dot{\zeta}^{(1)} = L_{\overline{f}}\mathbf{k}^1(x) + L_{\overline{g}}\mathbf{k}^1(x)u$$
$$\vdots$$
$$\dot{\zeta}^{(s)} = L_{\overline{f}}\mathbf{k}^s(x) + L_{\overline{g}}\mathbf{k}^s(x)u \qquad (3.39)$$

where the vector field $\overline{f}(x)$ and matrix $\overline{g}(x)$ are defined in Eq.3.34. Furthermore, the constraints $\mathbf{k}(x) = 0$ that specify the constrained state space $\mathcal{M}$, reduce to $\zeta_j^{(i)} = 0$, $i = 1,\ldots,s$; $j = 1,\ldots,(p-p_i)$. Thus, a direct elimination of the zero states $\zeta^{(i)}$, $i = 1,\ldots,s$, and the corresponding differential equations from Eq.3.39, yields the minimal-order state-space realization of Eq.3.36. □

Remark 3.2: In the coordinate change of Eq.3.35, the states $\zeta_i^{(0)}$, or equivalently the scalar functions $\phi_i(x)$, are chosen arbitrarily (they can be chosen simply as a proper subset $x_1, \ldots x_\kappa$ of the original differential variables x), and thus, the derivation of the state-space realization in Eq.3.36 requires an explicit solution for the algebraic variables z. However, in some cases (see Remark 7.3), it may be possible to choose the functions $\phi_i(x)$ in the coordinate change in a specific manner such that the differential equations for the corresponding state variables $\zeta_i^{(0)}$ do not involve the algebraic variables z, and thus, the minimal-order state-space realization can be derived without obtaining a solution for z. This is analogous to obtaining a representation of the zero dynamics of MIMO ODE systems independently of the solution for the manipulated inputs u, under appropriate involutivity conditions [64].

Remark 3.3: The algebraic constraints in Eq.3.24, identified by the algorithm, provide an explicit means for the choice of consistent initial conditions $x(0)$ for the differential variables, i.e., $\mathbf{k}(x(0)) = 0$. Thus, numerical simulation techniques for ODEs can be used for the solution of DAEs of the form in Eq.3.1, on the basis of the state-space realization in Eq.3.33 or the minimal-order realization in Eq.3.36. The latter representation of the system however, has the advantage of directly enforcing the underlying algebraic constraints in x at all times. In the simulation of the full-order representation of Eq.3.33, the solution for $x(t)$ may drift away from the constraints due to the presence of small numerical errors that accumulate over time.

In the following section, we will formulate and solve a feedback controller synthesis problem for regular DAE systems of the form in Eq.3.1.

3.4 State Feedback Controller Synthesis

3.4.1 Preliminaries

For a DAE system of Eq.3.1, various control-theoretic issues (such as existence and uniqueness of solutions, equilibrium points and their stability, zero dynamics, and characterization of minimum-phase behavior, etc.) can be addressed directly on the basis of the equivalent state-space realizations in Eq.3.33 or Eq.3.36, using existing results for ODE systems [64, 80, 108]. These state-space realizations can also be used as the basis for the formulation and solution of a variety of control problems for the DAE system, that are consistent with the underlying algebraic constraints in x. In what follows, we address the synthesis of state feedback controllers for stabilization and output tracking in the DAE system of Eq.3.1, within the general framework of input/output linearization (see also [87]). To this end, we initially introduce some basic concepts that are relevant to the analysis and controller synthesis, on the basis of the state-space realization of Eq.3.33.

For a DAE system of the form of Eq.3.1, the relative order (or degree) r_i of the output y_i, with respect to the manipulated input vector u, is defined as the minimum integer such that:

$$[L_{\bar{g}_1} L_f^{r_i-1} h_i(x) \cdots L_{\bar{g}_m} L_f^{r_i-1} h_i(x)] \neq [0 \cdots 0]$$

for $x \in X \subset \mathcal{M}$, where X is an open set containing the equilibrium point of interest. If no such integer exists, then $r_i = \infty$. It will be assumed that there is a finite relative order r_i for each output y_i, since this is necessary for output controllability. Then,

$$C(x) = \begin{bmatrix} L_{\bar{g}_1} L_f^{r_1-1} h_1(x) & \cdots & L_{\bar{g}_m} L_f^{r_1-1} h_1(x) \\ \vdots & & \vdots \\ L_{\bar{g}_1} L_f^{r_m-1} h_m(x) & \cdots & L_{\bar{g}_m} L_f^{r_m-1} h_m(x) \end{bmatrix} \tag{3.40}$$

denotes the characteristic (or decoupling) matrix for the system of Eq.3.33. For simplicity, it will be assumed that the characteristic matrix $C(x)$ (Eq.3.40) is nonsingular on X, allowing the synthesis of static state feedback controllers to induce a well-characterized linear input/output behavior in the closed-loop system. Moreover, it will be also assumed that the zero dynamics for the system of Eq.3.33 (or equivalently Eq.3.36) is locally asymptotically stable, i.e., the DAE system of Eq.3.1 is minimum-phase.

3.4.2 Problem formulation

Consider a DAE system of the form of Eq.3.1 with the equivalent state-space realization (Eq.3.33) and a nonsingular characteristic matrix $C(x)$. It is desired to derive a state feedback controller to enforce the following closed-loop objectives:

1. Induce a linear closed-loop input/output response of the form:

$$y + \sum_{i=1}^{m} \sum_{j=1}^{r_i} \gamma_{ij} \frac{d^j y_i}{dt^j} = y_{sp} \tag{3.41}$$

in the nominal system, where $y = [y_1 \cdots y_m]^T$, $y_{sp} = [y_{sp1} \cdots y_{spm}]^T$ are the output and setpoint vectors, and $\gamma_{ij} = [\gamma_{ij}^1 \cdots \gamma_{ij}^m]^T$ are vectors of adjustable parameters.

2. Asymptotically reject the effects of (constant) unmeasured disturbances and parametric errors in the models, on the controlled outputs.

3. Ensure closed-loop input/output and internal stability.

3.4.3 Controller synthesis

The controller synthesis problem for the DAE system of Eq.3.1 is addressed through a combination of a state feedback controller that induces a linear input/output behavior with respect to the inputs $v = [v_1 \cdots v_m]^T$:

$$\sum_{i=1}^{m} \sum_{j=0}^{r_i} \beta_{ij} \frac{d^j y_i}{dt^j} = v \tag{3.42}$$

where $\beta_{ij} = [\beta_{ij}^1 \cdots \beta_{ij}^m]^T$ are vectors of adjustable parameters, and a linear error feedback controller with integral action for the inputs v, to reject the effects of unmeasured disturbances and modeling errors. Theorem 3.1 that follows, states the result on the state feedback controller that induces the requested closed-loop input/output behavior in Eq.3.42.

Theorem 3.1: *Consider a DAE system of the form of Eq.3.1 with an equivalent state-space realization of Eq.3.33 for which* $\det C(x) \neq 0$, $\forall x \in X$. *Then, the state feedback law:*

$$u = \{[\beta_{1r_1} \cdots \beta_{mr_m}]C(x)\}^{-1} \left(v - \sum_{i=1}^{m} \sum_{j=0}^{r_i} \beta_{ij} L_f^j h_i(x) \right) \tag{3.43}$$

induces the input/output behavior:

$$\sum_{i=1}^{m} \sum_{j=0}^{r_i} \beta_{ij} \frac{d^j y_i}{dt^j} = v$$

subject to the underlying constraints imposed by the algebraic equations.

47

Proof: Consider the DAE system of Eq.3.1 which, under the control law of Eq.3.43, yields the following closed-loop DAE system:

$$\dot{x} = f(x) + b(x)z + g(x)\{[\beta_{1r_1} \cdots \beta_{mr_m}]C(x)\}^{-1}\left(v - \sum_{i=1}^{m}\sum_{j=0}^{r_i}\beta_{ij}L_{\bar{f}}^{j}h_i(x)\right)$$

$$0 = k(x) + l(x)z + c(x)\{[\beta_{1r_1} \cdots \beta_{mr_m}]C(x)\}^{-1}\left(v - \sum_{i=1}^{m}\sum_{j=0}^{r_i}\beta_{ij}L_{\bar{f}}^{j}h_i(x)\right) \qquad (3.44)$$

$$y_i = h_i(x); \ i = 1, \ldots, m$$

A state-space realization for the closed-loop DAE system of Eq.3.44 can be obtained following the algorithm proposed earlier. It is straightforward to show that the algorithm converges after exactly s iterations, identifying the same algebraic constraints in x (Eq.3.24) and yielding the following relation for z:

$$z = -\begin{bmatrix}\bar{l}^s(x) \\ \tilde{l}^{s+1}(x)\end{bmatrix}^{-1}\left\{\begin{bmatrix}\bar{k}^s(x) \\ \tilde{k}^{s+1}(x)\end{bmatrix} + \begin{bmatrix}\bar{c}^s(x) \\ \tilde{c}^{s+1}(x)\end{bmatrix}\right.$$

$$\left. \times \{[\beta_{1r_1} \cdots \beta_{mr_m}]C(x)\}^{-1}\left(v - \sum_{i=1}^{m}\sum_{j=0}^{r_i}\beta_{ij}L_{\bar{f}}^{j}h_i(x)\right)\right\} \qquad (3.45)$$

Thus, the state-space realization of the closed-loop dynamics takes the form:

$$\dot{x} = \bar{f}(x) + \bar{g}(x)C^{-1}(x)[\beta_{1r_1} \cdots \beta_{mr_m}]^{-1}\left\{v - \sum_{i=1}^{m}\sum_{j=0}^{r_i}\beta_{ij}L_{\bar{f}}^{j}h_i(x)\right\} \qquad (3.46)$$

$$y_i = h_i(x), \ i = 1, \ldots, m$$

where $x \in \mathcal{M}$ and $\bar{f}(x)$, $\bar{g}(x)$ are defined in Eq.3.34. Calculating the expressions for the derivatives of the outputs, i.e., $\dfrac{d^j y_i}{dt^j}, i = 1, \ldots, m; \ j = 1, \ldots, r_i$, on the basis of Eq.3.46 and substituting in Eq.3.42, it is then straightforward to show that the input/output behavior of Eq.3.42 is indeed enforced. □

The bounded-input bounded-output (BIBO) stability of the closed-loop system is ensured by a proper choice of the adjustable parameters β_{ij}^k. Besides BIBO stability of the closed-loop system, it is necessary to ensure the internal stability of the closed-loop system, i.e., the local asymptotic stability of the unforced ($v = 0$) closed-loop system. It can be verified that the unforced closed-loop system is locally asymptotically stable if the following two conditions hold:

1. The parameters β_{ij}^k are chosen properly to ensure BIBO stability of the system with the input/output behavior of Eq.3.42.

2. The zero dynamics of the process is locally asymptotically stable, i.e., the process is minimum-phase.

Given the state feedback controller of Eq.3.43 which induces the linear input/output behavior of Eq.3.42, a linear error feedback controller with integral action is incorporated around the linear $v-y$ system to enforce the requested closed-loop input/output behavior of Eq.3.41 in the nominal process and guarantee the asymptotic rejection of disturbances and modeling errors. The requisite error feedback controller has the following realization [39]:

$$\dot{\xi}_1^{(1)} = \xi_2^{(1)}$$

$$\vdots$$

$$\dot{\xi}_{r_1-1}^{(1)} = \xi_{r_1}^{(1)}$$

$$\dot{\xi}_{r_1}^{(1)} = ([\gamma_{1r_1} \cdots \gamma_{mr_m}]^{-1})_1 \left[(y_{sp} - y) - \sum_{i=1}^{m} \sum_{j=1}^{r_i-1} \gamma_{ij} \xi_{j+1}^{(i)} \right]$$

$$\vdots$$

$$\dot{\xi}_1^{(m)} = \xi_2^{(m)}$$

$$\vdots$$

$$\dot{\xi}_{r_m-1}^{(m)} = \xi_{r_m}^{(m)}$$

$$\dot{\xi}_{r_m}^{(m)} = ([\gamma_{1r_1} \cdots \gamma_{mr_m}]^{-1})_m \left[(y_{sp} - y) - \sum_{i=1}^{m} \sum_{j=1}^{r_i-1} \gamma_{ij} \xi_{j+1}^{(i)} \right]$$

$$v = \sum_{i=1}^{m} \sum_{j=0}^{r_i-1} \beta_{ij} \xi_{j+1}^{(i)} + [\beta_{1r_1} \cdots \beta_{mr_m}][\gamma_{1r_1} \cdots \gamma_{mr_m}]^{-1}[(y_{sp} - y) - \sum_{i=1}^{m} \sum_{j=1}^{r_i-1} \gamma_{ij} \xi_{j+1}^{(i)}]$$

$$(3.47)$$

where the symbol $()_i$ denotes ith row of a matrix.

3.5 Conclusions

In this chapter, we addressed the feedback control of a broad class of nonlinear MIMO high-index DAE systems in semiexplicit form through a systematic two-step procedure where in the first step, a state-space realization of the DAE system is derived, and subsequently used in the second step as the basis for the formulation and solution of a feedback controller synthesis problem. Owing to its sequential nature, this two-step approach is feasible only for the class of regular DAE systems. Initially, an algorithm was presented which allows a precise characterization of the class of DAE systems that are regular and yields state-space realizations of such systems. The derived state-space realization was then used for a state feedback controller synthesis to induce a well-characterized closed-loop input/output behavior with stability, subject to the underlying constraints imposed by the singular algebraic equations.

Notation

Roman letters

$$
\begin{aligned}
b, g, l, c &= \text{matrices in the DAE system} \\
C &= \text{characteristic matrix} \\
E^i &= p \times p \text{ nonsingular matrices} \\
f, k &= \text{vector fields in the DAE system} \\
\overline{f} &= \text{vector field in state-space realization of dimension } n \\
\overline{g} &= \text{matrix in state-space realization of dimension } n \\
\overline{k}^i, \mathbf{k}^i, \widetilde{k}^i &= \text{vector fields in the algorithm} \\
\overline{l}^i, \widetilde{l}^i, \overline{c}^i, \widetilde{c}^i &= \text{matrices in the algorithm} \\
h_i &= \text{scalar functions in the DAE system} \\
\mathcal{M} &= \text{constrained manifold where differential variables evolve} \\
m &= \text{number of manipulated inputs and controlled outputs} \\
n &= \text{number of differential variables} \\
p &= \text{number of algebraic variables} \\
p_i &= \text{ranks of matrices in the algorithm} \\
r_i &= \text{relative order of output } y_i \text{ with respect to } u \\
s &= \text{number of iterations for convergence of algorithm} \\
u &= \text{manipulated input vector} \\
v &= \text{external input vector} \\
x &= \text{vector of differential variables} \\
y_i &= \text{output variable} \\
y_{sp} &= \text{vector of output setpoints} \\
z &= \text{vector of algebraic variables}
\end{aligned}
$$

Greek letters

$$
\begin{aligned}
\beta_{ij}, \gamma_{ij} &= \text{vectors of adjustable controller parameters} \\
\zeta &= \text{vector of state variables in transformed coordinates} \\
\nu_d &= \text{index of a DAE system} \\
\xi &= \text{vector of states of linear error feedback controller} \\
\phi_i &= \text{scalar functions for coordinate transformation}
\end{aligned}
$$

Math symbols

$$
\begin{array}{rcl}
\mathbb{R} & = & \text{real line} \\
\mathbb{R}^i & = & i\text{-dimensional Euclidean space} \\
[\,\cdot\,]^T & = & \text{transpose of a vector/matrix} \\
L_f \alpha(x) & = & \text{Lie derivative of a scalar function } \alpha(x) \text{ with respect to} \\
\end{array}
$$

vector field $f(x)$, defined as $L_f \alpha(x) = [\dfrac{\partial \alpha}{\partial x_1} \; \cdots \; \dfrac{\partial \alpha}{\partial x_n}] f(x)$

$L_f^j \alpha(x)$ = high-order Lie derivative defined as $L_f^j \alpha(x) = L_f(L_f^{j-1} \alpha(x))$

4. Feedback Control of Nonregular DAE Systems

4.1 Introduction

In the previous chapter, we presented a two-step controller design methodology for the class of regular high-index DAE systems (for definition, see Section 1.3.4), for which a state-space realization could be derived in the first step and subsequently used for the controller synthesis in the second step. In this chapter, we address the feedback control of nonregular DAE systems, for which the underlying algebraic constraints in the differential variables explicitly involve the manipulated inputs, and thus, the constrained state space is control-dependent. For nonregular systems, the derivation of state-space realizations is coupled to the controller design, and thus, the sequential controller design approach of the previous chapter is not feasible. In this chapter, we discuss a controller design methodology for nonregular DAE systems developed in [88]. The key step in the controller design involves the design of a dynamic state feedback compensator to modify the DAE system such that in the resulting system, the underlying algebraic constraints in the differential variables are independent of the *new* inputs. Thus, the feedback compensator yields a modified DAE system that is regular. For this feedback regularized system, a state-space realization is derived and subsequently used as the basis for a state feedback controller synthesis.

4.2 Preliminaries

We consider multi-input multi-output (MIMO) nonlinear DAE systems with the description:

$$\dot{x} = f(x) + b(x)z + g(x)u$$
$$0 = k(x) + l(x)z + c(x)u$$
$$y_i = h_i(x), \quad i = 1, \ldots, m \tag{4.1}$$

where $x \in \mathcal{X} \subset \mathbb{R}^n$ is the vector of differential variables, $z \in \mathcal{Z} \subset \mathbb{R}^p$ is the vector of algebraic variables, $u \in \mathbb{R}^m$ is the vector of manipulated inputs and y_i is the ith output to be controlled. We focus on high-index systems of Eq.4.1 (i.e., the matrix $l(x)$ is singular on $\mathcal{X}$) for which the underlying algebraic constraints in x involve the manipulated inputs u, i.e., the regularity conditions of Lemma 3.1 are violated. For such nonregular systems, clearly, the constrained state space depends on the control law for u, and a state-space realization can not be derived independently of the controller design. However, the presence of the manipulated inputs allows modifying the nonregular DAE system, in particular the underlying algebraic constraints in x, in a desired manner.

The above facts motivate the following controller design methodology for nonregular DAE systems of Eq.4.1:

1. In the first step, a state feedback compensator is designed for the DAE system of Eq.4.1 such that the resulting modified system with a new vector of inputs is regular. To this end, initially a modified algorithm is developed to obtain a new DAE system, equivalent to the system in Eq.4.1, where the algebraic equations explicitly include the constraints in x that involve the inputs u, thereby isolating the cause of nonregularity. The design of the feedback regularizing compensator is then addressed on the basis of this equivalent DAE system.

2. In the second step, for the feedback regularized DAE system, state-space realizations are derived and used as the basis for the synthesis of a state feedback controller that induces a well-characterized input/output behavior. The resulting controller and the feedback regularizing compensator of Step 1 comprise the overall dynamic state feedback controller for the DAE system of Eq.4.1.

4.3 Algorithm for Derivation of Equivalent DAE System with Explicit Constraints in x Involving u

In this section, we modify the algorithm of Section 3.3.1 to account for the presence of the manipulated inputs u in the underlying algebraic constraints in x. The modified algorithm involves, in each iteration, (a) elementary row operations on the algebraic equations to identify the underlying constraints in x, of which a *minimal* number involve the inputs u, and (b) selective differentiation of only those constraints that do not involve u, to obtain the algebraic equations for the next iteration. The algorithm, thus avoids introducing any derivatives of the inputs u, and finally yields an equivalent DAE system for which a feedback regularizing compensator can be designed.

Iteration 1:

Consider the algebraic equations of the DAE system in Eq.4.1:

$$0 = k(x) + l(x)z + c(x)u \qquad (4.2)$$

where rank $l(x) = p_1 < p$ and rank $[l(x) \quad c(x)] = m_1 \leq p$ $(m_1 \geq p_1)$ on $\mathcal{X}$. Again we will assume that the region $\mathcal{X}$ is appropriately defined such that the ranks of the concerned matrices are constant on $\mathcal{X}$. There exists a smooth nonsingular $p \times p$ matrix

$E^1(x)$ such that:

$$E^1(x)\,[\,l(x)\ \ c(x)\,] = \begin{bmatrix} \bar{l}^1(x) & \bar{c}^1(x) \\ 0 & \hat{c}^1(x) \\ 0 & 0 \end{bmatrix} \qquad (4.3)$$

where $\bar{c}^1(x), \hat{c}^1(x)$ are matrices of dimensions $p_1 \times m$, $(m_1 - p_1) \times m$, respectively, and the $p_1 \times p$, $m_1 \times (p+m)$ matrices

$$\bar{l}^1(x), \qquad \begin{bmatrix} \bar{l}^1(x) & \bar{c}^1(x) \\ 0 & \hat{c}^1(x) \end{bmatrix}$$

have full row rank.

Step 1. Premultiply the algebraic equations (Eq.4.2) by the matrix $E^1(x)$ to obtain:

$$0 = \begin{bmatrix} \bar{k}^1(x) \\ \hat{k}^1(x) \\ \mathbf{k}^1(x) \end{bmatrix} + \begin{bmatrix} \bar{l}^1(x) \\ 0 \\ 0 \end{bmatrix} z + \begin{bmatrix} \bar{c}^1(x) \\ \hat{c}^1(x) \\ 0 \end{bmatrix} u \qquad (4.4)$$

where $\bar{k}^1(x), \hat{k}^1(x), \mathbf{k}^1(x)$ are smooth vector fields of dimensions $p_1, (m_1 - p_1)$ and $(p - m_1)$, respectively. The last $p - p_1$ equations in Eq.4.4 denote underlying constraints in x; a minimal number of these constraints (the first $m_1 - p_1$) involve the inputs u in an irreducible fashion, i.e., $\hat{c}^1(x)$ has full row rank.

Step 2. Differentiate the last $p - m_1$ constraints that do not involve u, to obtain the following new set of algebraic equations:

$$0 = \begin{bmatrix} \bar{k}^1(x) \\ \hat{k}^1(x) \\ \tilde{k}^2(x) \end{bmatrix} + \begin{bmatrix} \bar{l}^1(x) \\ 0 \\ \tilde{l}^2(x) \end{bmatrix} z + \begin{bmatrix} \bar{c}^1(x) \\ \hat{c}^1(x) \\ \tilde{c}^2(x) \end{bmatrix} u \qquad (4.5)$$

In the above equation, $\tilde{k}^2(x) = [L_f \mathbf{k}_1^1(x) \ \cdots \ L_f \mathbf{k}_{p-m_1}^1(x)]^T$ where $\mathbf{k}_i^1(x)$ denotes the ith component of the vector field $\mathbf{k}^1(x)$, and $\tilde{l}^2(x), \tilde{c}^2(x)$ are matrices of dimensions $(p - m_1) \times p$ and $(p - m_1) \times m$, respectively, defined as:

$$\tilde{l}^2(x) = \begin{bmatrix} L_{b_1}\mathbf{k}_1^1(x) & \cdots & L_{b_p}\mathbf{k}_1^1(x) \\ \vdots & & \vdots \\ L_{b_1}\mathbf{k}_{p-m_1}^1(x) & \cdots & L_{b_p}\mathbf{k}_{p-m_1}^1(x) \end{bmatrix}, \quad \tilde{c}^2(x) = \begin{bmatrix} L_{g_1}\mathbf{k}_1^1(x) & \cdots & L_{g_m}\mathbf{k}_1^1(x) \\ \vdots & & \vdots \\ L_{g_1}\mathbf{k}_{p-m_1}^1(x) & \cdots & L_{g_m}\mathbf{k}_{p-m_1}^1(x) \end{bmatrix}$$

where $b_j(x), g_j(x)$ denote the jth columns of corresponding matrices, and $L_f \mathbf{k}_i^1(x)$, $L_{b_j}\mathbf{k}_i^1(x), L_{g_j}\mathbf{k}_i^1(x)$ denote standard Lie derivatives.

Step 3. Evaluate:

$$\text{rank}\begin{bmatrix} \bar{l}^1(x) \\ \tilde{l}^2(x) \end{bmatrix} = p_2, \qquad \text{rank}\begin{bmatrix} \bar{l}^1(x) & \bar{c}^1(x) \\ 0 & \hat{c}^1(x) \\ \tilde{l}^2(x) & \tilde{c}^2(x) \end{bmatrix} = m_2, \quad m_2 \geq p_2$$

If $m_2 = p$ then stop, else proceed to the next iteration. For the case when $m_1 = p$, the algorithm converges after Step 1, and all the $p - p_1$ constraints in x, i.e., $0 = \widehat{k}^1(x) + \widehat{c}^1(x)u$, involve u in an irreducible fashion.

Iteration q $(q \geq 2)$:

Consider the algebraic equations from iteration $q - 1$:

$$0 = \begin{bmatrix} \overline{k}^{q-1}(x) \\ \widehat{k}^{q-1}(x) \\ \widetilde{k}^q(x) \end{bmatrix} + \begin{bmatrix} \overline{l}^{q-1}(x) \\ 0 \\ \widetilde{l}^q(x) \end{bmatrix} z + \begin{bmatrix} \overline{c}^{q-1}(x) \\ \widehat{c}^{q-1}(x) \\ \widetilde{c}^q(x) \end{bmatrix} u \qquad (4.6)$$

where the matrices:

$$L^q(x) = \begin{bmatrix} \overline{l}^{q-1}(x) \\ \widetilde{l}^q(x) \end{bmatrix}, \qquad L^{q,e}(x) = \begin{bmatrix} \overline{l}^{q-1}(x) & \overline{c}^{q-1}(x) \\ 0 & \widehat{c}^{q-1}(x) \\ \widetilde{l}^q(x) & \widetilde{c}^q(x) \end{bmatrix}$$

have ranks p_q, m_q, respectively $(p_q \leq m_q < p)$. Then, there exists a nonsingular $p \times p$ matrix $E^{q,1}(x)$ of the form:

$$E^{q,1}(x) = \begin{bmatrix} I_{m_{q-1}} & 0 \\ R^q(x) & S^q(x) \end{bmatrix}$$

such that:

$$E^{q,1}(x)L^{q,e}(x) = \begin{bmatrix} \overline{l}^{q-1}(x) & \overline{c}^{q-1}(x) \\ 0 & \widehat{c}^{q-1}(x) \\ \cdots\cdots\cdots\cdots \\ l^q(x) & c^{q,1}(x) \\ 0 & c^{q,2}(x) \\ 0 & 0 \end{bmatrix}$$

where $l^q(x)$, $c^{q,1}(x)$ and $c^{q,2}(x)$ are matrices of dimensions $(p_q - p_{q-1}) \times p$, $(p_q - p_{q-1}) \times m$, and $(m_q - m_{q-1} - p_q + p_{q-1}) \times m$, respectively. The last $p - m_q$ rows of the above matrix are identically zero. Furthermore, there exists a $p \times p$ permutation matrix $E^{q,2}$ which rearranges the rows of $E^{q,1}(x)L^{q,e}(x)$ to obtain:

$$E^{q,2}E^{q,1}(x)L^{q,e}(x) = \begin{bmatrix} \overline{l}^q(x) & \overline{c}^q(x) \\ 0 & \widehat{c}^q(x) \\ 0 & 0 \end{bmatrix}$$

where:

$$\overline{c}^q(x) = \begin{bmatrix} \overline{c}^{q-1}(x) \\ c^{q,1}(x) \end{bmatrix}, \qquad \widehat{c}^q(x) = \begin{bmatrix} \widehat{c}^{q-1}(x) \\ c^{q,2}(x) \end{bmatrix}$$

and the $p_q \times p$, $m_q \times (p+m)$ matrices:

$$\overline{l}^q(x) = \begin{bmatrix} \overline{l}^{q-1}(x) \\ l^q(x) \end{bmatrix}, \qquad \begin{bmatrix} \overline{l}^q(x) & \overline{c}^q(x) \\ 0 & \widehat{c}^q(x) \end{bmatrix}$$

have full row rank. Define $E^q(x) = E^{q,2} E^{q,1}(x)$.

Step 1. Premultiply the algebraic equations (Eq.4.6) with the matrix $E^q(x)$ to obtain:

$$0 = \begin{bmatrix} \overline{k}^q(x) \\ \widehat{k}^q(x) \\ k^q(x) \end{bmatrix} + \begin{bmatrix} \overline{l}^q(x) \\ 0 \\ 0 \end{bmatrix} z + \begin{bmatrix} \overline{c}^q(x) \\ \widehat{c}^q(x) \\ 0 \end{bmatrix} u \qquad (4.7)$$

where $\overline{k}^q(x), \widehat{k}^q(x), k^q(x)$ are smooth vector fields of dimensions $p_q, (m_q - p_q), (p - m_q)$, respectively, and the last $p - p_q$ equations denote underlying algebraic constraints in x.

Step 2. Differentiate the constraints in x that are independent of u, i.e., the last $p - m_q$ equations in Eq.4.7, to obtain the following new set of algebraic equations:

$$0 = \begin{bmatrix} \overline{k}^q(x) \\ \widehat{k}^q(x) \\ \widetilde{k}^{q+1}(x) \end{bmatrix} + \begin{bmatrix} \overline{l}^q(x) \\ 0 \\ \widetilde{l}^{q+1}(x) \end{bmatrix} z + \begin{bmatrix} \overline{c}^q(x) \\ \widehat{c}^q(x) \\ \widetilde{c}^{q+1}(x) \end{bmatrix} u \qquad (4.8)$$

where $\widetilde{k}^{q+1}(x)$ is a vector field of dimension $p - m_q$ and $\widetilde{l}^{q+1}(x)$, $\widetilde{c}^{q+1}(x)$ are matrices of dimensions $(p - m_q) \times p$, $(p - m_q) \times m$, respectively, defined in a fashion analogous to that in Iteration 1.

Step 3. Evaluate the rank of the following matrices:

$$\text{rank} \begin{bmatrix} \overline{l}^q(x) \\ \widetilde{l}^{q+1}(x) \end{bmatrix} = p_{q+1}, \qquad \text{rank} \begin{bmatrix} \overline{l}^q(x) & \overline{c}^q(x) \\ 0 & \widehat{c}^q(x) \\ \widetilde{l}^{q+1}(x) & \widetilde{c}^{q+1}(x) \end{bmatrix} = m_{q+1}, \quad m_{q+1} \geq p_{q+1}$$

If $m_{q+1} = p$ then stop, else repeat the above steps for the next iteration, starting with the algebraic equations in Eq.4.8.

For a DAE system of Eq.4.1 with a finite index ν_d, the above algorithm will converge in a finite number of iterations s with $p_1 \leq p_2 \leq \cdots \leq p_{s+1} < p$ and $m_2 \leq m_3 \leq \cdots \leq m_{s+1} = p$, $(m_i \geq p_i, \forall i > 1)$. The algorithm identifies the following algebraic constraints in the differential variables x:

$$\mathbf{k}(x) = \begin{bmatrix} \mathbf{k}^1(x) \\ \vdots \\ \mathbf{k}^s(x) \end{bmatrix} = 0 \qquad (4.9)$$

which do not involve the inputs u. Similar to the result in Proposition 3.1, it can be shown that the constraints in Eq.4.9 are linearly independent, i.e., the gradient

56

covector fields $d\mathbf{k}_i(x) = [\frac{\partial \mathbf{k}_i}{\partial x_1}(x) \cdots \frac{\partial \mathbf{k}_i}{\partial x_n}(x)]$ are linearly independent, where $\mathbf{k}_i(x)$ is the i-th component of the vector field $\mathbf{k}(x)$. The algorithm also yields the following set of algebraic equations:

$$0 = \begin{bmatrix} \overline{k}^s(x) \\ \widehat{k}^s(x) \\ \widetilde{k}^{s+1}(x) \end{bmatrix} + \begin{bmatrix} \overline{l}^s(x) \\ 0 \\ \widetilde{l}^{s+1}(x) \end{bmatrix} z + \begin{bmatrix} \overline{c}^s(x) \\ \widehat{c}^s(x) \\ \widetilde{c}^{s+1}(x) \end{bmatrix} u \qquad (4.10)$$

where the matrices:

$$L^{s+1}(x) = \begin{bmatrix} \overline{l}^s(x) \\ \widetilde{l}^{s+1}(x) \end{bmatrix}, \qquad L^{s+1,e}(x) = \begin{bmatrix} \overline{l}^s(x) & \overline{c}^s(x) \\ 0 & \widehat{c}^s(x) \\ \widetilde{l}^{s+1}(x) & \widetilde{c}^{s+1}(x) \end{bmatrix}$$

have ranks p_{s+1} and $m_{s+1} = p$, respectively. Thus, there exists a nonsingular $p \times p$ matrix $E^{s+1}(x)$ such that:

$$E^{s+1}(x)L^{s+1,e}(x) = \begin{bmatrix} \overline{l}(x) & \overline{c}(x) \\ 0 & \widehat{c}(x) \end{bmatrix}$$

where $\overline{c}(x)$ and $\widehat{c}(x)$ are matrices of dimensions $p_{s+1} \times m$ and $(p-p_{s+1}) \times m$, respectively, and the $p_{s+1} \times p$, $p \times (p+m)$ matrices:

$$\overline{l}(x), \qquad \begin{bmatrix} \overline{l}(x) & \overline{c}(x) \\ 0 & \widehat{c}(x) \end{bmatrix}$$

have full row rank. Premultiplying the algebraic equations in Eq.4.10 with the matrix $E^{s+1}(x)$, the following final set of algebraic equations are obtained:

$$0 = \begin{bmatrix} \overline{k}(x) \\ \widehat{k}(x) \end{bmatrix} + \begin{bmatrix} \overline{l}(x) \\ 0 \end{bmatrix} z + \begin{bmatrix} \overline{c}(x) \\ \widehat{c}(x) \end{bmatrix} u \qquad (4.11)$$

where $\overline{k}(x)$ and $\widehat{k}(x)$ are smooth vector fields of dimensions p_{s+1} and $p - p_{s+1}$, respectively. Thus, the algorithm yields the following new DAE system:

$$\begin{aligned} \dot{x} &= f(x) + b(x)z + g(x)u \\ 0 &= \begin{bmatrix} \overline{k}(x) \\ \widehat{k}(x) \end{bmatrix} + \begin{bmatrix} \overline{l}(x) \\ 0 \end{bmatrix} z + \begin{bmatrix} \overline{c}(x) \\ \widehat{c}(x) \end{bmatrix} u \\ y_i &= h_i(x), \quad i = 1,\ldots,m \end{aligned} \qquad (4.12)$$

where $x \in \mathcal{X}$: $\mathbf{k}(x) = 0$. The above DAE system (Eq.4.12) is equivalent to the original DAE system in Eq.4.1, i.e., for consistent initial conditions $x(0)$ and smooth inputs $u(t)$, both systems have the same solution $x(t), z(t)$. Moreover, note that the algebraic equations in Eq.4.12 explicitly include the underlying constraints in x,

$0 = \widehat{k}(x) + \widehat{c}(x)u$, that involve the inputs u in an irreducible fashion, i.e., $\widehat{c}(x)$ has a full row rank $p - p_{s+1}$. In the next section, the nonregular DAE system of Eq.4.12 will be *regularized* through feedback, to facilitate the derivation of state-space realizations that can be used as the basis for feedback controller synthesis.

Remark 4.1: For a regular DAE system of the form in Eq.4.1, the algorithm reduces to that in section 3.3.1, and it converges after exactly $s = \nu_d - 1$ iterations. On the other hand, for nonregular DAE systems, the algorithm will converge in s iterations, where s may be significantly less than $\nu_d - 1$.

4.4 Dynamic State Feedback Regularization

Consider the nonregular DAE system in Eq.4.12, obtained through the algorithm. The aim is to design a state feedback compensator such that the resulting feedback modified DAE system with a new vector of inputs is regular. Note, however, that the DAE system in Eq.4.12 still has a high index $\overline{\nu}_d = \nu_d - s > 1$, and the constraints $0 = \widehat{k}(x) + \widehat{c}(x)u$ have to be differentiated at least once to obtain a set of algebraic equations solvable in z. Thus, the solution for the algebraic variables z is a function of the differential variables x, the manipulated inputs u and *at least* one of their derivatives, i.e., it has the form:

$$z(t) = \varphi(x(t), u(t), u^{(1)}(t), \ldots) \qquad (4.13)$$

where $u^{(i)}$ denotes the ith derivative of the manipulated input vector u. In light of this fact, any causal feedback law for u must clearly be independent of z. Furthermore, any (regular) static feedback of the form:

$$u = \mathcal{F}(x, v)$$

where $v \in \mathbb{R}^m$ is the new input vector and $(\partial \mathcal{F} / \partial v)$ is nonsingular, will not regularize the DAE system of Eq.4.12, since the constraints $0 = \widehat{k}(x) + \widehat{c}(x)\mathcal{F}(x, v)$ in the resulting modified system would still involve the inputs v.

The above observations indicate the need for a *dynamic* feedback compensator, to modify the DAE system in Eq.4.12, in particular, the constraints:

$$0 = \widehat{k}(x) + \widehat{c}(x)u \qquad (4.14)$$

that cause the nonregularity, such that the resulting system with the inputs v is regular. A systematic derivation of the compensator involves the following two steps.
Step 1. Employ an input transformation:

$$u = M(x) \begin{bmatrix} \overline{u}_1 \\ \overline{u}_2 \end{bmatrix}$$

where $\bar{u}_1 \in \mathbb{R}^{(p-p_{s+1})}$, $\bar{u}_2 \in \mathbb{R}^{m-(p-p_{s+1})}$, to isolate the $p - p_{s+1}$ inputs $\bar{u}_1$ that appear in the constraints of Eq.4.14 in a nonsingular fashion. This can be accomplished in a straightforward fashion by choosing $M(x)$ as a nonsingular $m \times m$ matrix such that:

$$\begin{bmatrix} \bar{c}(x) \\ \hat{c}(x) \end{bmatrix} M(x) = \begin{bmatrix} \bar{c}_1(x) & \bar{c}_2(x) \\ \hat{c}_1(x) & 0 \end{bmatrix} \tag{4.15}$$

where the $(p - p_{s+1}) \times (p - p_{s+1})$ matrix $\hat{c}_1(x)$ is nonsingular. Under the above transformation, the constraints in Eq.4.14 take the following form:

$$0 = \hat{k}(x) + \hat{c}_1(x)\bar{u}_1 \tag{4.16}$$

Step 2. Design a dynamic feedback compensator for the inputs $\bar{u}_1$ with the following general form:

$$\dot{w} = v_1$$
$$\bar{u}_1 = (\hat{c}_1(x))^{-1} \left[-\hat{k}(x) + C(x, w) \right]$$

where $w \in \mathbb{R}^{p-p_{s+1}}$ is the vector of compensator states and $C(x, w)$ is a smooth vector field of dimension $p - p_{s+1}$, to modify the constraints in Eq.4.16 so that:

(a) the resulting constraints:

$$0 = C(x, w) \tag{4.17}$$

involve only the differential variables x and the compensator states w, and

(b) the new set of algebraic equations obtained after one differentiation of the above constraints (Eq.4.17) can be solved for the algebraic variables z.

From the fact that the feedback compensator must be independent of z, it is evident that upon differentiation of the modified constraints (Eq.4.17) in Step 2b, the algebraic variables z can appear only through the differential variables x. Thus, the dynamic feedback compensator of Step 2, in particular $C(x, w)$, must be designed to alter the dependence of the constraints in Eq.4.17 on the variables x in a way such that the requirement in Step 2b is met. The existence of such a compensator is ensured by an important characteristic of semiexplicit DAE systems of the form in Eq.4.1 with a finite index ν_d, stated in the following lemma.

Lemma 4.1: *Consider a solvable semiexplicit DAE system described by:*

$$\dot{x} = f(x) + b(x)z + g(x)u(t)$$
$$0 = k(x) + l(x)z + c(x)u(t) \tag{4.18}$$

with a finite index ν_d, where $x \in \mathcal{X} \subset \mathbb{R}^n$, $z \in \mathcal{Z} \subset \mathbb{R}^p$, $u(t) = [u_1(t) \cdots u_m(t)]^T$ is a vector of smooth inputs, $f(x), k(x)$ are smooth vector fields of dimensions n, p, *respectively, and $b(x), g(x), l(x), c(x)$ are matrices of appropriate dimensions. Then, the $(n + p) \times p$ matrix:*

$$\begin{bmatrix} b(x) \\ l(x) \end{bmatrix} \tag{4.19}$$

has full column rank on $\mathcal{X}$.

Proof: Consider a solvable DAE system of the form of Eq.4.18, with a finite index $\nu_d > 1$ for a smooth input $u(t)$. For simplicity, consider $u(t) = \alpha$, where $\alpha \in \mathbb{R}^m$ is a constant vector. The corresponding DAE system is described by:

$$\begin{aligned} \dot{x} &= \widetilde{f}(x) + b(x)z \\ 0 &= \widetilde{k}(x) + l(x)z \end{aligned} \tag{4.20}$$

where $\widetilde{f}(x) = f(x) + g(x)\alpha$, $\widetilde{k}(x) = k(x) + c(x)\alpha$ and $\operatorname{rank} l(x) = p_1 < p$. Then, by the definition of index, a set of differential equations for z in the DAE system of Eq.4.20 can be obtained by differentiating a proper subset of the algebraic equations ν_d times. In particular, the algorithm of section 3.3.1 will converge in exactly $s = \nu_d - 1$ iterations to yield a final set of algebraic equations that are nonsingular with respect to the algebraic variables z.

We prove through contradiction that for the algorithm to converge, it is necessary that the matrix:

$$\begin{bmatrix} l(x) \\ b(x) \end{bmatrix}$$

has full column rank p. More specifically, assume that:

$$\operatorname{rank} \begin{bmatrix} l(x) \\ b(x) \end{bmatrix} = p^* < p \tag{4.21}$$

Then, through induction, it is shown that $p_i \leq p^* < p$ for every iteration i of the algorithm, hence, proving the assumption to be wrong.

Consider the assumption in Eq.4.21. It leads to the following two conclusions:

(a) For the choice of matrix $E^1(x)$ in Step 1 of the first iteration:

$$\operatorname{rank} \begin{bmatrix} b(x) \\ l^1(x) \\ 0 \end{bmatrix} = \operatorname{rank} \left\{ \begin{bmatrix} I_n & 0 \\ 0 & E^1(x) \end{bmatrix} \begin{bmatrix} b(x) \\ l(x) \end{bmatrix} \right\} \leq p^* \tag{4.22}$$

i.e.,

$$\operatorname{rank} \begin{bmatrix} \bar{l}^1(x) \\ b(x) \end{bmatrix} \leq p^* < p$$

(b) If, for any iteration $i \geq 1$,

$$\text{rank} \begin{bmatrix} \bar{l}^i(x) \\ b(x) \end{bmatrix} \leq p^* < p$$

then,

(i) from Step 2 of the algorithm:

$$\text{rank} \begin{bmatrix} \bar{l}^i(x) \\ \tilde{l}^{i+1}(x) \end{bmatrix} = \text{rank} \left\{ \begin{bmatrix} I_{p_i} & 0 \\ 0 & \dfrac{d\hat{k}^i(x)}{dx} \end{bmatrix} \begin{bmatrix} \bar{l}^i(x) \\ b(x) \end{bmatrix} \right\} \leq p^*$$

i.e.,

$$\text{rank} \begin{bmatrix} \bar{l}^i(x) \\ \tilde{l}^{i+1}(x) \end{bmatrix} = p_{i+1} \leq p^* < p$$

(ii) consider the matrix:

$$E^{i+1}(x) = \begin{bmatrix} E_1^{i+1}(x) \\ E_2^{i+1}(x) \end{bmatrix}$$

where $E_1^{i+1}(x)$ is the matrix consisting of the first p_{i+1} rows of $E^{i+1}(x)$, i.e.,

$$E_1^{i+1}(x) \begin{bmatrix} \bar{l}^i(x) \\ \tilde{l}^{i+1}(x) \end{bmatrix} = \bar{l}^{i+1}(x)$$

Thus,

$$\text{rank} \begin{bmatrix} \bar{l}^{i+1}(x) \\ \cdots \\ b(x) \end{bmatrix} = \text{rank} \left\{ \begin{bmatrix} E^{i+1}(x) & 0 \\ \cdots \\ 0 & I_n \end{bmatrix} \begin{bmatrix} I_{p_i} & 0 \\ 0 & \dfrac{d\hat{k}^i(x)}{dx} \\ \cdots \\ 0 & I_n \end{bmatrix} \begin{bmatrix} \bar{l}^i(x) \\ b(x) \end{bmatrix} \right\} \leq p^* < p$$

Clearly, from the above two conclusions (a) and (b), it follows that $p_i \leq p^* < p$ for every iteration $i \geq 1$ and the algorithm will not converge, thus yielding a contradiction for the assumption in Eq.4.21.   □

Remark 4.2: Consider the following linear analogue of the DAE system in Eq.4.18:

$$\begin{aligned} \dot{x} &= Ax + Bz + Gu(t) \\ 0 &= Kx + Lz + Cu(t) \end{aligned} \tag{4.23}$$

where A, B, G, K, L, C are constant matrices of appropriate dimensions. Then, for a solvable DAE system of Eq.4.23 with a finite index $\nu_d < n$, the result of Lemma 4.1

follows directly [89] from the fact that the matrix pencil:

$$\mathcal{P} = \begin{bmatrix} \lambda I_n - A & -B \\ -K & -L \end{bmatrix}$$

is regular, i.e., $\det \mathcal{P} \neq 0$ for some $\lambda \in \mathbb{C}$.

The result of Lemma 4.1 implies that for the DAE system in Eq.4.12, the matrix:

$$\begin{bmatrix} b(x) \\ \bar{l}(x) \end{bmatrix}$$

has full rank p, and thus, there exists a matrix $S \in \mathbb{R}^{(p-p_s+1) \times n}$ such that the following $p \times p$ matrix:

$$\begin{bmatrix} Sb(x) \\ \bar{l}(x) \end{bmatrix} \tag{4.24}$$

is nonsingular. This fact implies that the desired goal of feedback regularization will be achieved if the constraint in Eq.4.17, obtained through feedback modification, has the form:

$$0 = \mathcal{C}(x, w) = Sx + w \tag{4.25}$$

Theorem 4.1 that follows, provides a representation of the dynamic feedback regularizing compensator derived along the above lines.

Theorem 4.1: *Consider a DAE system of the form in Eq.4.1, for which the algorithm yields the equivalent DAE system of Eq.4.12. Then, the dynamic feedback compensator:*

$$\dot{w} = v_1$$
$$u = M(x) \begin{bmatrix} (\widehat{c}_1(x))^{-1} \left(-\widehat{k}(x) + Sx + w \right) \\ 0 \end{bmatrix} + M(x) \begin{bmatrix} 0 \\ v_2 \end{bmatrix} \tag{4.26}$$

where w, $v_1 \in \mathbb{R}^{(p-p_s+1)}$ and $v_2 \in \mathbb{R}^{m-(p-p_s+1)}$, and the matrices S, $M(x)$ are chosen as in Eq.4.24,4.15, respectively, yields the modified DAE system:

$$\begin{bmatrix} \dot{x} \\ \dot{w} \end{bmatrix} = \begin{bmatrix} \widetilde{f}(x, w) \\ 0 \end{bmatrix} + \begin{bmatrix} b(x) \\ 0 \end{bmatrix} z + \begin{bmatrix} 0 & \bar{g}_2(x) \\ I_{p-p_s+1} & 0 \end{bmatrix} \begin{bmatrix} v_1 \\ v_2 \end{bmatrix}$$
$$0 = \begin{bmatrix} \widetilde{k}(x, w) \\ Sx + w \end{bmatrix} + \begin{bmatrix} \bar{l}(x) \\ 0 \end{bmatrix} z + \begin{bmatrix} 0 & \bar{c}_2(x) \\ 0 & 0 \end{bmatrix} \begin{bmatrix} v_1 \\ v_2 \end{bmatrix} \tag{4.27}$$
$$y_i = h_i(x), \quad i = 1, \ldots, m$$

where:

$$\widetilde{f}(x, w) = f(x) + \bar{g}_1(x)\gamma(x) + \bar{g}_1(x)(\widehat{c}_1(x))^{-1} w$$

62

$$\widetilde{k}(x, w) = \overline{k}(x) + \overline{c}_1(x)\gamma(x) + \overline{c}_1(x)(\widehat{c}_1(x))^{-1}w$$
$$\gamma(x) = (\widehat{c}_1(x))^{-1}\{-\widehat{k}(x) + Sx\}$$
$$[\overline{g}_1(x) \quad \overline{g}_2(x)] = g(x)M(x) \tag{4.28}$$

and $x \in \mathcal{X}$: $\mathbf{k}(x) = 0$. *The DAE system of Eq.4.27 with the extended vector of differential variables $\bar{x} = [x^T \ w^T]^T$ and a new vector of inputs $v = [v_1^T \ v_2^T]^T \in \mathbb{R}^m$ is regular.*

Proof: For the DAE system in Eq.4.12, the dynamic feedback compensator of Eq.4.26 directly yields the extended DAE system in Eq.4.27, where the last $p - p_{s+1}$ algebraic equations, $Sx + w = 0$, denote constraints in the differential variables $\bar{x}$ that are independent of the inputs v. Differentiating these constraints once, the following set of algebraic equations is obtained:

$$0 = \begin{bmatrix} \widetilde{k}(x, w) \\ S\widetilde{f}(x, w) \end{bmatrix} + \begin{bmatrix} \overline{l}(x) \\ Sb(x) \end{bmatrix} z + \begin{bmatrix} 0 & \overline{c}_2(x) \\ I_{p-p_{s+1}} & S\overline{g}_2(x) \end{bmatrix} \begin{bmatrix} v_1 \\ v_2 \end{bmatrix} \tag{4.29}$$

Clearly, by the choice of the matrix S as in Eq.4.24, the coefficient matrix for z in the above algebraic equations, i.e.:

$$\begin{bmatrix} \overline{l}(x) \\ Sb(x) \end{bmatrix}$$

is nonsingular. Thus, the DAE system in Eq.4.27 is solvable, with index $\nu_d = 2$ and the unique smooth solution for z:

$$z = -\begin{bmatrix} \overline{l}(x) \\ Sb(x) \end{bmatrix}^{-1} \left\{ \begin{bmatrix} \widetilde{k}(x, w) \\ S\widetilde{f}(x, w) \end{bmatrix} + \begin{bmatrix} 0 & \overline{c}_2(x) \\ I_{p-p_{s+1}} & S\overline{g}_2(x) \end{bmatrix} \begin{bmatrix} v_1 \\ v_2 \end{bmatrix} \right\}$$
$$= R(x, w) + S_1(x)v_1 + S_2(x)v_2 \tag{4.30}$$

Moreover, the fact that Eq.4.29 is solvable in z, implies that there are no additional constraints in the differential variables $\bar{x}$, and the constraints $\mathbf{k}(x) = 0$, $Sx + w = 0$, which are independent of the inputs v, specify the $n - \sum\limits_{i=1}^{s}(p - m_i)$-dimensional subspace:

$$\mathcal{M} = \left\{ (x, w) \in \mathcal{X} \times \mathbb{R}^{(p-p_{s+1})} \ : \ \begin{matrix} \mathbf{k}(x) = 0 \\ Sx + w = 0 \end{matrix} \right\} \subset \mathbb{R}^{n+(p-p_{s+1})} \tag{4.31}$$

It can be verified that for any initial condition $\bar{x}(0) \in \mathcal{M}$ and under the unique solution of z in Eq.4.30, the differential variables $\bar{x}(t)$ are constrained to evolve in the subspace $\mathcal{M}$ for any smooth input $v(t)$. Thus, $\mathcal{M}$ is the constrained state space of the DAE system in Eq.4.27, which is invariant under any control law for the inputs v. This establishes that the feedback modified DAE system of Eq.4.27 is regular. $\square$

Remark 4.3: The choice of the constant matrix S in Eq.4.24 is not unique. For the purpose of regularization, S can always be chosen to be a simple permutation matrix that selects $p - p_{s+1}$ rows of the matrix $b(x)$ which, together with the matrix $\bar{l}(x)$, comprise a nonsingular $p \times p$ matrix. On the other hand, one has the flexibility to choose a more general matrix S, which affects the constrained state space $\mathcal{M}$ and hence, the state-space realization of the feedback regularized DAE system, to attain additional objectives besides regularization. For a specific choice of the matrix S, the initial condition $w(0) \in \mathbb{R}^{p-p_{s+1}}$ for the compensator states can be chosen to be consistent with the constraints $0 = Sx(0) + w(0)$.

A schematic representation of the overall feedback regularization procedure for a nonregular DAE system of Eq.4.1 is shown in Figure 4.1.

4.5 State-Space Realizations and Controller Synthesis

For the feedback regularized DAE system in Eq.4.27, the specification of the constrained state space $\mathcal{M}$ (Eq.4.31) and the solution for z (Eq.4.30) allow the derivation of a state-space realization given in Proposition 4.1 that follows.

Proposition 4.1: *Consider a DAE system of the form in Eq.4.1, for which the algorithm converges after s iterations and the dynamic feedback compensator of Theorem 4.1 yields the regularized extended DAE system in Eq.4.27. Then, the dynamic system:*

$$\dot{\bar{x}} = \overline{f}(\bar{x}) + \overline{g}(\bar{x})v$$
$$y_i = \overline{h}_i(\bar{x}), \quad i = 1, \dots, m \tag{4.32}$$

is a state-space realization of the feedback regularized DAE system, where $\bar{x} = [x^T \ w^T]^T \in \mathcal{M}$ is the extended state vector, $v \in \mathbb{R}^m$ is the new input vector defined in Theorem 4.1, and

$$\overline{f}(\bar{x}) = \begin{bmatrix} \tilde{f}(x,w) + b(x)R(x,w) \\ 0 \end{bmatrix}, \quad \overline{g}(\bar{x}) = \begin{bmatrix} b(x)S_1(x) & \overline{g}_2(x) + b(x)S_2(x) \\ I_{p-p_{s+1}} & 0 \end{bmatrix}$$

$$\overline{h}_i(\bar{x}) = h_i(x) \tag{4.33}$$

Proof: Consider the feedback regularized extended DAE system in Eq.4.27. In Theorem 4.1, it was established that for consistent initial conditions $\bar{x}(0) \in \mathcal{M}$, the solution $\bar{x}(t)$ of Eq.4.27 is constrained to evolve in $\mathcal{M}$ for all times, with the corresponding solution for z in Eq.4.30. A direct substitution of the solution for z in the differential equations for $\bar{x}$ yields the differential equations in Eq.4.32 on the constrained state

64

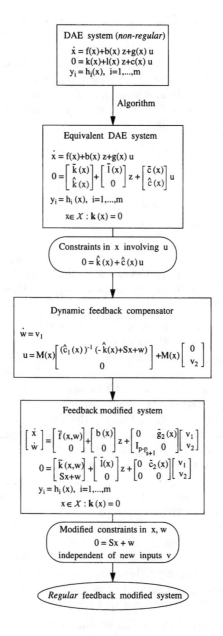

Figure 4.1: Steps in feedback regularization of a nonregular DAE system

space $\mathcal{M}$, for which the solution $\bar{x}(t)$ is the same as that for the DAE system in Eq.4.27. □

In view of the dimension $\kappa = n - \sum_{i=1}^{s}(p - m_i)$ of the constrained state space $\mathcal{M}$ (Eq.4.31) for the state variables $\bar{x}$, the state-space realization in Eq.4.32 is clearly not of minimal order. A minimal-order realization can be obtained in a suitably transformed set of coordinates. More specifically, using a nonlinear coordinate transformation of the form:

$$\zeta = \begin{bmatrix} \zeta_1^{(0)} \\ \vdots \\ \zeta_\kappa^{(0)} \\ \zeta^{(1)} \\ \zeta^{(2)} \end{bmatrix} = T(\bar{x}) = \begin{bmatrix} \phi_1(x) \\ \vdots \\ \phi_\kappa(x) \\ \mathbf{k}(x) \\ Sx + w \end{bmatrix} \tag{4.34}$$

where $\zeta^{(0)} \in \mathbb{R}^\kappa$, $\zeta^{(1)} \in \mathbb{R}^{n-\kappa}$, $\zeta^{(2)} \in \mathbb{R}^{p-p_{s+1}}$ and $\phi_1(x), \ldots, \phi_\kappa(x)$ are scalar functions chosen such that $\zeta = T(\bar{x})$ in Eq.4.34 is a local diffeomorphism and eliminating the states $\zeta^{(1)}, \zeta^{(2)}$ that are identically zero on the state space $\mathcal{M}$, a state-space realization of dimension κ can be obtained.

Remark 4.4: For a class of DAE systems where the algebraic constraints in the differential variables x involve the manipulated inputs u, an approach based on dynamic extension has been proposed in [137]. In this approach, *all* the constraints in x are differentiated a sufficient number of times ($\nu_d - 1$ to be precise) until a set of equations is obtained that can be solved for the algebraic variables z in terms of x, the inputs u, and the derivatives of these inputs. The inputs u and their derivatives that appear in the constraints in x, are included in the state vector to obtain a state-space realization of an extended DAE system. However, in general, this state-space realization and the corresponding dynamic feedback controller are of inordinately high order. Moreover, the approach is applicable to systems for which the index ν_d is well-defined and constant, irrespectively of the nature of the (time-varying) inputs u. This is quite restrictive, since the index of a general nonregular DAE system with time-varying inputs may vary (see Example 1.2 in Section 1.3.3). In contrast, the proposed approach involves a selective differentiation of the constraints in x that do not involve u, thereby avoiding any derivatives of u, and does not require the index ν_d to be constant for any arbitrary u; we require only that the DAE system has a finite index ν_d for *some* smoothly time-varying input $u(t)$. Furthermore, the resulting dynamic feedback compensator in Eq.4.26 is exactly of the same order ($p - p_{s+1}$) as the minimal number of constraints in x that involve the inputs u.

A state feedback controller synthesis problem for the DAE system of Eq.4.1 can now be formulated and solved on the basis of the state-space realization (Eq.4.32) of the feedback regularized system (Eq.4.27). To this end, the notions of equilibrium points, stability of solutions, zero dynamics and characterization of minimum-phase behavior, and relative orders between the outputs and inputs can be introduced for the regularized DAE system in Eq.4.27, on the basis of the state-space realization of Eq.4.32. Specifically, for the feedback regularized DAE system of Eq.4.27, the relative order r_i of the controlled output y_i, with respect to the manipulated input vector v, is defined as the minimum integer such that:

$$\left[L_{\bar{g}_1} L_{\bar{f}}^{r_i-1} \bar{h}_i(\bar{x}) \quad L_{\bar{g}_2} L_{\bar{f}}^{r_i-1} \bar{h}_i(\bar{x}) \quad \cdots \quad L_{\bar{g}_m} L_{\bar{f}}^{r_i-1} \bar{h}_i(\bar{x}) \right] \not\equiv [0 \quad 0 \quad \cdots \quad 0]$$

for $\bar{x} \in X \subset \mathcal{M}$, where X is an open connected set containing the equilibrium point of interest. It is assumed that a finite relative order r_i exists for every controlled output y_i, since it is necessary for output controllability. The characteristic matrix:

$$C(\bar{x}) = \begin{bmatrix} L_{\bar{g}_1} L_{\bar{f}}^{r_1-1} \bar{h}_1(\bar{x}) & \cdots & L_{\bar{g}_m} L_{\bar{f}}^{r_1-1} \bar{h}_1(\bar{x}) \\ \vdots & & \vdots \\ L_{\bar{g}_1} L_{\bar{f}}^{r_m-1} \bar{h}_m(\bar{x}) & \cdots & L_{\bar{g}_m} L_{\bar{f}}^{r_m-1} \bar{h}_m(\bar{x}) \end{bmatrix} \tag{4.35}$$

will be also assumed to be nonsingular on X, for simplicity. Finally, it is assumed that the zero dynamics for the system of Eq.4.32 is locally asymptotically stable, i.e., the system is minimum-phase.

Consider the DAE system of Eq.4.1, for which the proposed regularization scheme yields the regular extended DAE system of Eq.4.27 with the state-space realization of Eq.4.32, a nonsingular characteristic matrix $C(x)$ and a locally asymptotically stable zero dynamics. For such a system, it is desired to derive a state feedback controller that enforces the following closed-loop characteristics:

1. Induce an input/output behavior described by:

$$\sum_{i=1}^{m} \sum_{j=0}^{r_i} \beta_{ij} \frac{d^j y_i}{dt^j} = \tilde{v} \tag{4.36}$$

where $\tilde{v} = [\tilde{v}_1 \quad \cdots \quad \tilde{v}_m]^T$ is the vector of external reference inputs and $\beta_{ij} = [\beta_{ij}^1 \quad \cdots \quad \beta_{ij}^m]^T$ are vectors of adjustable parameters.

2. Ensure input/output and internal stability.

A solution to the above synthesis problem will be derived on the basis of the state-space realization (Eq.4.32) of the feedback regularized DAE system (Eq.4.27). More specifically, a static state feedback controller will be derived for the system in Eq.4.32 to induce the input/output behavior of Eq.4.36 in the closed-loop system. The resulting controller, together with the dynamic feedback regularizing compensator of Theorem 4.1, will comprise the overall *dynamic* state feedback controller for the DAE system of Eq.4.1, as shown in Figure 4.2. The main result of this section is stated in the following theorem.

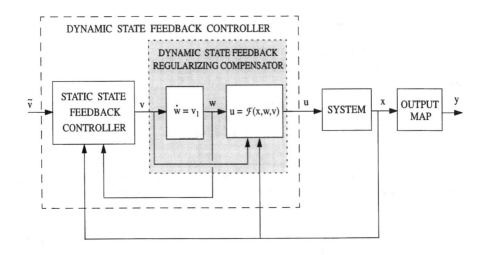

Figure 4.2: Overall structure of dynamic state feedback controller

Theorem 4.2: *Consider a DAE system of the form of Eq.4.1, for which the proposed algorithm and dynamic feedback regularizing compensator of Theorem 4.1 yield the regular extended DAE system of Eq.4.27 with the state-space realization of Eq.4.32 and a characteristic matrix $C(\bar{x})$, where $\det C(\bar{x}) \neq 0$, $\forall \ \bar{x} \in X$. Then, for the DAE system in Eq.4.1, the dynamic state feedback controller:*

$$\dot{w} = v_1$$

$$\begin{bmatrix} v_1 \\ v_2 \end{bmatrix} = \{[\beta_{1r_1} \cdots \beta_{mr_m}]C(x,w)\}^{-1} \left(\widetilde{v} - \sum_{i=1}^{m}\sum_{j=0}^{r_i} \beta_{ij} L_{\overline{\mathcal{F}}}^{j} \overline{h}_i(x,w) \right) \tag{4.37}$$

$$u = M(x)\begin{bmatrix} (\widehat{c}_1(x))^{-1}\left(-\widehat{k}(x) + Sx + w\right) \\ 0 \end{bmatrix} + M(x)\begin{bmatrix} 0 \\ v_2 \end{bmatrix}$$

induces the input/output behavior:

$$\sum_{i=1}^{m}\sum_{j=0}^{r_i} \beta_{ij}\frac{d^j y_i}{dt^j} = \widetilde{v}$$

subject to the algebraic constraints that describe the constrained subspace in Eq.4.31.

Proof: Consider the DAE system in Eq.4.1 and the dynamic state feedback control law of Eq.4.37. The proposed algorithm yields the DAE system of Eq.4.12, which is equivalent to the DAE system in Eq.4.1, in the sense that both systems have the same solutions $x(t)$, $z(t)$ and hence, the same outputs $y_i(t)$, for *any* (smooth) input $u(t)$. Thus, the closed-loop DAE system of Eq.4.1 under the control law of Eq.4.37 is equivalent (in the above sense) to the closed-loop DAE system of Eq.4.12 under the same control law (Eq.4.37), and it suffices to prove the result for the latter system.

First, consider the DAE system of Eq.4.12 under the control law for the manipulated inputs u in terms of the new inputs $v = [v_1^T \ v_2^T]^T$ in Eq.4.26, i.e., the extended DAE system in Eq.4.27. Differentiating the last $p-p_{s+1}$ algebraic equations in Eq.4.27 once, we obtain the equivalent DAE system:

$$\begin{bmatrix} \dot{x} \\ \dot{w} \end{bmatrix} = \begin{bmatrix} \tilde{f}(x,w) \\ 0 \end{bmatrix} + \begin{bmatrix} b(x) \\ 0 \end{bmatrix} z + \begin{bmatrix} 0 & \bar{g}_2(x) \\ I_{p-p_{s+1}} & 0 \end{bmatrix} \begin{bmatrix} v_1 \\ v_2 \end{bmatrix}$$

$$0 = \begin{bmatrix} \tilde{k}(x,w) \\ S\tilde{f}(x,w) \end{bmatrix} + \begin{bmatrix} \bar{l}(x) \\ Sb(x) \end{bmatrix} z + \begin{bmatrix} 0 & \bar{c}_2(x) \\ I_{p-p_{s+1}} & S\bar{g}_2(x) \end{bmatrix} \begin{bmatrix} v_1 \\ v_2 \end{bmatrix} \qquad (4.38)$$

$$y_i = h_i(x), \quad i = 1, \dots, m$$

$\bar{x} = [x^T \ w^T]^T \in \mathcal{M}$, for *any* inputs v. Moreover, from Proposition 4.1, the above DAE system (Eq.4.38) has an equivalent state-space realization of Eq.4.32 for *any* input v, and in particular, the specific control law given by:

$$v = \{[\beta_{1r_1} \cdots \beta_{mr_m}]C(\bar{x})\}^{-1} \left(\widetilde{v} - \sum_{i=1}^{m}\sum_{j=0}^{r_i} \beta_{ij} L_f^j \bar{h}_i(\bar{x}) \right)$$

Thus, the DAE system of Eq.4.1 under the overall dynamic state-feedback control law of Eq.4.37 is equivalent to the ODE system described by:

$$\dot{\bar{x}} = \bar{f}(\bar{x}) + \bar{g}(\bar{x})\{[\beta_{1r_1} \cdots \beta_{mr_m}]C(\bar{x})\}^{-1} \left(\widetilde{v} - \sum_{i=1}^{m}\sum_{j=0}^{r_i} \beta_{ij} L_f^j \bar{h}_i(\bar{x}) \right)$$

$$y_i = \bar{h}_i(\bar{x}), \quad i = 1, \dots, m \qquad (4.39)$$

on the constrained state space $\mathcal{M}$. For the above system, by obtaining the expressions for the derivatives of the outputs, i.e., $\dfrac{d^j y_i}{dt^j}$, $i = 1, \dots, m$; $j = 1, \dots, r_i$, it can be verified that the desired input/output response in Eq.4.36 is indeed enforced in the closed-loop system. $\qquad \square$

Under a proper choice of the adjustable parameters β_{ij}^k that ensure the bounded-input bounded-output (BIBO) stability of the input/output behavior described in Eq.4.36, and the assumption of local asymptotic stability of the zero dynamics of the system in Eq.4.32, it can be verified that the closed-loop system is stable. Furthermore, a suitable linear error feedback controller with integral action is incorporated around the linearized $\widetilde{v} - y$ system to induce the following closed-loop input/output behavior in the nominal system

$$y + \sum_{i=1}^{m}\sum_{j=1}^{r_i} \gamma_{ij}\frac{d^j y_i}{dt^j} = y_{sp} \qquad (4.40)$$

and asymptotically reject the effects of small parametric modeling errors [39]. In the input/output response requested in Eq.4.40, $y = [y_1 \cdots y_m]^T$, $y_{sp} = [y_{sp1} \cdots y_{spm}]^T$ are the output and setpoint vectors, while $\gamma_{ij} = [\gamma_{ij}^1 \cdots \gamma_{ij}^m]^T$ are vectors of adjustable parameters.

4.6 Conclusions

In this chapter, we addressed the feedback control of nonregular, high-index DAE systems of the form in Eq.4.1, for which the underlying constraints in the differential variables involve the manipulated inputs. In contrast with regular DAE systems for which the underlying constraints in the differential variables are independent of the inputs, a state-space realization of nonregular systems can not be derived independently of the control law for the manipulated inputs. Motivated by this, a control methodology was presented where the key step involved a feedback modification of the DAE system such that the resulting system is regular. To this end, initially, the algorithm of the previous chapter was suitably modified to identify the underlying constraints and obtain an equivalent system where the algebraic equations explicitly include the constraints involving the manipulated inputs, thereby isolating the cause of nonregularity. A feedback regularizing compensator was then designed for this equivalent DAE system. Finally, a state-space realization of the feedback regularized system was derived and used as the basis for the synthesis of a dynamic state feedback controller that induces a well-characterized, linear input/output behavior in the closed-loop system.

Notation

Roman letters

b, g, l, c = matrices in the DAE system
C = characteristic matrix
E^i = nonsingular matrices in the algorithm
f, k = vector fields in the DAE system
$\overline{k}^i, \widehat{k}^i, \widetilde{k}^i, \mathbf{k}^i$ = vector fields in the algorithm
$\overline{l}^i, \widetilde{l}^i, \overline{c}^i, \widehat{c}^i, \widetilde{c}^i$ = matrices in the algorithm
h_i = scalar functions in the DAE system
M = matrix used in feedback regularizing compensator
m = number of manipulated inputs and controlled outputs
n = number of differential variables
p = number of algebraic variables
p_i, m_i = ranks of matrices in the algorithm
r_i = relative order of output y_i with respect to u
S = constant matrix in feedback regularizing compensator
s = number of iterations for convergence of algorithm
u = manipulated input vector
v = input vector for extended DAE system
$\widetilde{v}$ = external input vector
w = state of the dynamic feedback regularizing compensator
$x, \bar{x}$ = vector of differential variables for original, and extended DAE systems, respectively.
y_i = output variable
y_{sp} = vector of output setpoints
z = vector of algebraic variables

Greek letters

β_{ij}, γ_{ij} = vectors of adjustable controller parameters
ξ = state of the linear error feedback controller
ζ = vector of state variables in transformed coordinates
ν_d = index of a DAE system
ϕ_i = scalar functions in coordinate transformation

Math symbols

$\mathcal{M}$	=	constrained manifold where differential variables of the extended DAE system evolve
$\mathbb{R}$	=	real line
$\mathbb{R}^i$	=	i-dimensional Euclidean space
$\mathbb{C}$	=	set of complex numbers
$[\,\cdot\,]^T$	=	transpose of a vector/matrix
$L_f\alpha(x)$	=	Lie derivative of a scalar function $\alpha(x)$ with respect to a vector field $f(x)$, defined as $L_f\alpha(x) = [\dfrac{\partial \alpha}{\partial x_1} \;\cdots\; \dfrac{\partial \alpha}{\partial x_n}]f(x)$
$L_f^j\alpha(x)$	=	high-order Lie derivative defined as $L_f^j\alpha(x) = L_f(L_f^{j-1}\alpha(x))$

5. Control of DAE Systems with Disturbance Inputs

5.1 Introduction

In this chapter, we address the control of DAE systems in the presence of time-varying disturbance inputs. It is well-known that for nonlinear ODE systems, the presence of disturbance inputs, especially time-varying ones, may lead to significant performance deterioration if they are not accounted for in the controller design. This fact motivated research on eliminating the effect of unmeasured disturbances on the controlled outputs through static state feedback [62, 65, 135], or if disturbance measurements are available, through static feedforward/state feedback [105] (see also [64, 108]) and dynamic feedforward/static state feedback [38, 41] controllers.

For a nonlinear high-index DAE system, the presence of time-varying disturbance inputs may have fundamental consequences on the system characteristics and the controller design. More specifically, similar to the case with the manipulated inputs, the disturbance inputs may also appear explicitly in the underlying algebraic constraints. In such a case, owing to the arbitrary time-varying nature of these disturbance inputs, the DAE system may fail to even have a well-defined index (see Section 1.3.3) and the derived state-space realizations for regular or feedback regularized DAE systems may not be well-defined.

Motivated by the above observations, we will identify and focus on the class of regular DAE systems for which the underlying algebraic constraints do not involve the disturbance inputs. For such systems, a well-defined state-space realization exists irrespectively of the time-varying disturbances [84, 85]. On the other hand, we will also identify the class of nonregular DAE systems for which the disturbances can be eliminated from the algebraic constraints through a feedforward/feedback modification, and a well-defined state-space realization can be derived for the resulting regularized system. Finally, the synthesis of feedforward/feedback controllers for these classes of regular/nonregular DAE systems will be addressed on the basis of the derived state-space realizations.

5.2   System Description and Preliminaries

We consider nonlinear DAE systems with the following description:

$$\dot{x} = f(x) + b(x)z + g(x)u + \alpha(x)d(t)$$
$$0 = k(x) + l(x)z + c(x)u \qquad\qquad (5.1)$$
$$y_i = h_i(x), \quad i = 1, \ldots, m$$

where the effect of the disturbance inputs $d(t) \in \mathbb{R}^\nu$ is explicitly modeled. In the above description, $x \in \mathcal{X} \subset \mathbb{R}^n$ is the vector of differential variables, $z \in \mathcal{Z} \subset \mathbb{R}^p$ is the vector of algebraic variables, $u \in \mathbb{R}^m$ is the vector of manipulated inputs, and y_i is the ith output to be controlled. Note that the disturbance inputs $d(t)$ appear in an affine fashion, only in the differential equations. This is motivated by the fact that typical disturbances in inlet feed flow rates, composition, temperature, etc., appear explicitly only in the dynamic conservation equations for mass and energy. The disturbance inputs $d(t)$ are assumed to be measured and smoothly time-varying.

In analogy with the approach for the feedback control of DAE systems without disturbances, the controller design for the DAE system in Eq.5.1 will be addressed on the basis of suitable state-space realizations. In the previous chapters, DAE systems were categorized into two broad classes depending on the nature of the underlying algebraic constraints in x: (i) regular systems for which the constraints in x do not involve the manipulated inputs u, and thus, a well-defined state-space realization exists independently of the controller design, and (ii) nonregular systems for which the constraints in x explicitly involve u. For the latter class of systems, the index depends on the input $u(t)$ and may not be well-defined (see Section 1.3.3). Furthermore, a state-space realization of such systems does not exist independently of the controller design. However, the fact that the inputs u can be manipulated as desired allowed the design of a feedback compensator such that the resulting feedback modified system possessed a well-defined index and state-space realization.

Similar to the manipulated inputs u, the disturbance inputs $d(t)$ may also appear explicitly in the underlying algebraic constraints in x. The presence of these disturbances (which vary with time in an arbitrary "uncontrolled" manner) in the algebraic constraints will, in general, lead to problems in the existence of well-defined index and state-space realizations. Motivated by this, we will focus on the class of DAE systems for which a well-defined state-space realization exists, or if the system is nonregular, a state-space realization of a regularized system exists, despite the disturbances $d(t)$. More specifically, we will focus on the class of regular DAE systems for which the algebraic constraints in x are independent of d. On the other hand, in the case of nonregular systems, we will focus on systems for which a feedforward/feedback compensator can be used to eliminate the disturbances from the algebraic constraints and obtain a regular system with a well-defined state-space realization. In the rest of this chapter, we will often drop the argument t from the disturbances $d(t)$ with the understanding that they are always smoothly time-varying.

5.3 Algorithm for Identification of Constraints

For DAE systems without disturbances, the algorithm of Section 4.3 was developed to identify the underlying algebraic constraints in x, separating the ones that do not involve the manipulated inputs u from those that do. The algorithm reduces to the one in Section 3.3.1 for regular systems. In this section, we will modify the algorithm to account for the disturbances d (see also [85] for a variant of this algorithm). Note that the algebraic equations in the DAE system of Eq.5.1 do not involve the disturbances. Thus, for rank $l(x) = p_1 < p$ and rank $[l(x)\ c(x)] = m_1 \leq p$ on $\mathcal{X}$, the first iteration proceeds the same as in Section 4.3 to identify the $p - m_1$ constraints, $0 = \mathbf{k}^1(x)$, that are independent of u and yield the following set of algebraic equations:

$$0 = \begin{bmatrix} \overline{k}^1(x) \\ \widehat{k}^1(x) \\ \widetilde{k}^2(x) \end{bmatrix} + \begin{bmatrix} \overline{l}^1(x) \\ 0 \\ \widetilde{l}^2(x) \end{bmatrix} z + \begin{bmatrix} \overline{c}^1(x) \\ \widehat{c}^1(x) \\ \widetilde{c}^2(x) \end{bmatrix} u + \begin{bmatrix} 0 \\ 0 \\ \widetilde{\beta}^2(x) \end{bmatrix} d \qquad (5.2)$$

where $\widetilde{\beta}^2(x)$ is a $(p - m_1) \times \nu$ matrix defined in an analogous fashion to $\widetilde{l}^2(x)$ and $\widetilde{c}^2(x)$. Note that in the above set of algebraic equations, the last $p - m_1$ equations explicitly involve the disturbance inputs d, and thus, these disturbances may appear in the constraints in x, in the succeeding iterations.

Iteration q $(q \geq 2)$:

Consider the algebraic equations from iteration $q - 1$:

$$0 = \begin{bmatrix} \overline{k}^{q-1}(x) \\ \widehat{k}^{q-1}(x) \\ \widetilde{k}^q(x) \end{bmatrix} + \begin{bmatrix} \overline{l}^{q-1}(x) \\ 0 \\ \widetilde{l}^q(x) \end{bmatrix} z + \begin{bmatrix} \overline{c}^{q-1}(x) \\ \widehat{c}^{q-1}(x) \\ \widetilde{c}^q(x) \end{bmatrix} u + \begin{bmatrix} \overline{\beta}^{q-1}(x) \\ \widehat{\beta}^{q-1}(x) \\ \widetilde{\beta}^q(x) \end{bmatrix} d \qquad (5.3)$$

where the matrices:

$$L^q(x) = \begin{bmatrix} \overline{l}^{q-1}(x) \\ \widetilde{l}^q(x) \end{bmatrix}, \quad L^{q,e}(x) = \begin{bmatrix} \overline{l}^{q-1}(x) & \overline{c}^{q-1}(x) \\ 0 & \widehat{c}^{q-1}(x) \\ \widetilde{l}^q(x) & \widetilde{c}^q(x) \end{bmatrix}$$

$$L^{q,eI}(x) = \begin{bmatrix} \overline{l}^{q-1}(x) & \overline{\beta}^{q-1}(x) \\ 0 & \widehat{\beta}^{q-1}(x) \\ \widetilde{l}^q(x) & \widetilde{\beta}^q(x) \end{bmatrix} \text{ and } L^{q,eII}(x) = \begin{bmatrix} \overline{l}^{q-1}(x) & \overline{c}^{q-1}(x) & \overline{\beta}^{q-1}(x) \\ 0 & \widehat{c}^{q-1}(x) & \widehat{\beta}^{q-1}(x) \\ \widetilde{l}^q(x) & \widetilde{c}^q(x) & \widetilde{\beta}^q(x) \end{bmatrix}$$

$$(5.4)$$

have ranks $p_q, m_q, \nu_{q,1}$ and $\nu_{q,2}$ on $\mathcal{X}$, respectively $(p_q \leq m_q < p)$. The above set of algebraic equations in Eq.5.3 imposes $p - p_q$ constraints in x. Depending on the ranks m_q and $\nu_{q,1}$, these constraints may or may not involve the manipulated inputs u and disturbances d. More specifically, for regular systems, $m_q = p_q$ and the constraints are independent of the manipulated inputs u. We will consider regular systems for

75

which a well-defined state-space realization exists irrespectively of the time-varying disturbances $d(t)$, i.e., systems for which the constraints in x are also independent of d. Clearly, this is true, if and only if, $\nu_{q,1} = p_q$. Similarly, we will consider nonregular DAE systems of Eq.5.1 for which the disturbances can be eliminated from the constraints in x through a feedforward/feedback compensation. Clearly, this is possible if, and only if, the disturbances d appear only in those constraints in x which also involve the manipulated inputs u in a nonsingular fashion. It can be easily verified that a necessary and sufficient condition for this to be true is $\nu_{q,2} = m_q$.

Under the condition $\nu_{q,2} = m_q$, a nonsingular $p \times p$ matrix $E^q(x) = E^{q,2}E^{q,1}(x)$ can be chosen in a similar manner as in the algorithm in Section 4.3, such that:

$$E^q(x)L^{q,eII}(x) = \begin{bmatrix} \bar{l}^q(x) & \bar{c}^q(x) & \bar{\beta}^q(x) \\ 0 & \hat{c}^q(x) & \hat{\beta}^q(x) \\ 0 & 0 & 0 \end{bmatrix}$$

where the $p_q \times p$ and $m_q \times (p+m)$ matrices:

$$\bar{l}^q(x), \quad \begin{bmatrix} \bar{l}^q(x) & \bar{c}^q(x) \\ 0 & \hat{c}^q(x) \end{bmatrix}$$

have full row rank.

Step 1. Premultiply the algebraic equations in Eq.5.3 with the matrix $E^q(x)$ to obtain:

$$0 = \begin{bmatrix} \bar{k}^q(x) \\ \hat{k}^q(x) \\ \mathbf{k}^q(x) \end{bmatrix} + \begin{bmatrix} \bar{l}^q(x) \\ 0 \\ 0 \end{bmatrix} z + \begin{bmatrix} \bar{c}^q(x) \\ \hat{c}^q(x) \\ 0 \end{bmatrix} u + \begin{bmatrix} \bar{\beta}^q(x) \\ \hat{\beta}^q(x) \\ 0 \end{bmatrix} d \qquad (5.5)$$

The last $p - p_q$ equations in Eq.5.5 denote constraints in x, of which the first $m_q - p_q$ involve the manipulated inputs u and possibly the disturbances d, while the last $p - m_q$ are independent of both u and d.

Step 2. Differentiate the last $p - m_q$ equations of Eq.5.5, i.e., the constraints in x that are independent of the manipulated inputs u and the disturbances d, to obtain the following set of algebraic equations:

$$0 = \begin{bmatrix} \bar{k}^q(x) \\ \hat{k}^q(x) \\ \tilde{k}^{q+1}(x) \end{bmatrix} + \begin{bmatrix} \bar{l}^q(x) \\ 0 \\ \tilde{l}^{q+1}(x) \end{bmatrix} z + \begin{bmatrix} \bar{c}^q(x) \\ \hat{c}^q(x) \\ \tilde{c}^{q+1}(x) \end{bmatrix} u + \begin{bmatrix} \bar{\beta}^q(x) \\ \hat{\beta}^q(x) \\ \tilde{\beta}^{q+1}(x) \end{bmatrix} d \qquad (5.6)$$

where $\tilde{k}^{q+1}(x)$ is a vector field with the ith component $\tilde{k}_i^{q+1}(x) = L_f \mathbf{k}_i^q(x)$, and $\tilde{l}^{q+1}(x), \tilde{c}^{q+1}(x)$ and $\tilde{\beta}^{q+1}(x)$ are matrices with the (i,j)th elements $\tilde{l}_{i,j}^{q+1}(x) = L_{b_j} \mathbf{k}_i^q(x)$, $\tilde{c}_{i,j}^{q+1}(x) = L_{g_j} \mathbf{k}_i^q(x)$, $\tilde{\beta}_{i,j}^{q+1}(x) = L_{\alpha_j} \mathbf{k}_i^q(x)$; $b_j(x), g_j(x)$ and $\alpha_j(x)$ denote the jth column vectors of the respective matrices.

Step 3. For the new set of algebraic equations in Eq.5.6, evaluate the ranks p_{q+1}, m_{q+1}, $\nu_{q+1,1}$ and $\nu_{q+1,2}$ of the matrices:

$$L^{q+1}(x) = \begin{bmatrix} \overline{l}^q(x) \\ \widetilde{l}^{q+1}(x) \end{bmatrix}, \quad L^{q+1,e}(x) = \begin{bmatrix} \overline{l}^q(x) & \overline{c}^q(x) \\ 0 & \widehat{c}^q(x) \\ \widetilde{l}^{q+1}(x) & \widetilde{c}^{q+1}(x) \end{bmatrix}$$

$$L^{q+1,eI}(x) = \begin{bmatrix} \overline{l}^q(x) & \overline{\beta}^q(x) \\ 0 & \widehat{\beta}^q(x) \\ \widetilde{l}^{q+1}(x) & \widetilde{\beta}^{q+1}(x) \end{bmatrix} \quad \text{and} \quad L^{q+1,eII}(x) = \begin{bmatrix} \overline{l}^q(x) & \overline{c}^q(x) & \overline{\beta}^q(x) \\ 0 & \widehat{c}^q(x) & \widehat{\beta}^q(x) \\ \widetilde{l}^{q+1}(x) & \widetilde{c}^{q+1}(x) & \widetilde{\beta}^{q+1}(x) \end{bmatrix}$$

$$(5.7)$$

respectively. If $m_{q+1}=p$ then stop, else proceed to the next iteration.

5.4 State-Space Realizations and Feedforward/State Feedback Control of Regular Systems

For a regular DAE system of Eq.5.1, $m_q=p_q$, $\forall q \geq 1$, and the algorithm converges in exactly $s=\nu_d-1$ iterations, identifying the $\sum_{i=1}^{s}(p-p_i)$ constraints in the differential variables x:

$$\mathbf{k}(x) = \begin{bmatrix} \mathbf{k}^1(x) \\ \vdots \\ \mathbf{k}^s(x) \end{bmatrix} = 0 \tag{5.8}$$

which are independent of the manipulated inputs u and the disturbances d and yielding the following final set of algebraic equations:

$$0 = \begin{bmatrix} \overline{k}^s(x) \\ \widetilde{k}^{s+1}(x) \end{bmatrix} + \begin{bmatrix} \overline{l}^s(x) \\ \widetilde{l}^{s+1}(x) \end{bmatrix} z + \begin{bmatrix} \overline{c}^s(x) \\ \widetilde{c}^{s+1}(x) \end{bmatrix} u + \begin{bmatrix} \overline{\beta}^s(x) \\ \widetilde{\beta}^{s+1}(x) \end{bmatrix} d \tag{5.9}$$

where:

$$\text{rank} \begin{bmatrix} \overline{l}^s(x) \\ \widetilde{l}^{s+1}(x) \end{bmatrix} = p$$

The above set of equations can be solved for the algebraic variables:

$$z = - \begin{bmatrix} \overline{l}^s(x) \\ \widetilde{l}^{s+1}(x) \end{bmatrix}^{-1} \left\{ \begin{bmatrix} \overline{k}^s(x) \\ \widetilde{k}^{s+1}(x) \end{bmatrix} + \begin{bmatrix} \overline{c}^s(x) \\ \widetilde{c}^{s+1}(x) \end{bmatrix} u + \begin{bmatrix} \overline{\beta}^s(x) \\ \widetilde{\beta}^{s+1}(x) \end{bmatrix} d \right\}$$

$$= a_1(x) + a_2(x)u + a_3(x)d \tag{5.10}$$

Note that the solution for the algebraic variables z is well-defined for *any* time-varying disturbances $d(t)$. This is true because the algebraic constraints in x (Eq.5.8) are independent of the disturbances d. The following lemma formally states the result on the class of regular DAE systems in Eq.5.1 for which, the underlying algebraic constraints are independent of the disturbances d and a well-defined state-space realization exists in the presence of arbitrary time-varying disturbances.

Lemma 5.1: *Consider a regular DAE system of Eq.5.1, i.e., $m_i = p_i$ for all $i \geq 1$. The underlying algebraic constraints in the differential variables x in the system are independent of the disturbances, if and only if, the condition $\nu_{q,1} = p_q$ holds for every iteration $q \geq 2$ of the algorithm.*

Proof: Consider the algebraic equations in Eq.5.3 in iteration q for a DAE system of Eq.5.1 that is regular, i.e., $m_q = p_q$. These equations impose $p - p_q$ algebraic constraints in x. It is straightforward to verify that the condition $\nu_{q,1} = p_q$ is necessary and sufficient for these constraints to be independent of the disturbances d. Furthermore, the condition $\nu_{q,1} = p_q$ is independent of the choice of the matrix $E^q(x)$ in the successive iterations and is an intrinsic property of the DAE system. This follows from the fact that the integer ranks p_q, m_q, and $\nu_{q,1}$ are not affected by the choice of $E^q(x)$, which can be established following a procedure similar to [61]. Thus, for the DAE system of Eq.5.1, for any specific set of matrices $E^q(x)$ in the algorithm, the condition $\nu_{q,1} = p_q$ for $q \geq 2$ is necessary and sufficient for the $\sum_{i=1}^{s}(p - p_i)$ constraints in x to be independent of the disturbances d. This completes the proof. □

For a regular DAE system of Eq.5.1, satisfying the condition $\nu_{q,1} = p_q$ for $q \geq 2$, the solution for z in Eq.5.10 and the specification of the subspace:

$$\mathcal{M} = \{x \in \mathcal{X} \subset \mathbb{R}^n \ : \ \mathbf{k}(x) = 0\} \tag{5.11}$$

where the differential variables x are constrained to evolve, allow the derivation of state-space realizations irrespectively of the disturbances. A state-space realization of dimension n is given in the following proposition.

Proposition 5.1: *Consider a DAE system of Eq.5.1 for which the algorithm converges after s iterations, with $m_i = p_i$ for $i \geq 1$ and $\nu_{i,1} = p_i$ for $i \geq 2$. Then, the dynamic system:*

$$\begin{aligned}
\dot{x} &= \overline{f}(x) + \overline{g}(x)u + \overline{\alpha}(x)d \\
y_i &= h_i(x), \quad i = 1, \ldots, m
\end{aligned} \tag{5.12}$$

where $x \in \mathcal{M} = \{x \in \mathcal{X} \ : \ \mathbf{k}(x) = 0\}$,

$$\overline{f}(x) = f(x) + b(x)a_1(x), \quad \overline{g}(x) = g(x) + b(x)a_2(x), \quad \overline{\alpha}(x) = \alpha(x) + b(x)a_3(x) \tag{5.13}$$

and $a_1(x), a_2(x), a_3(x)$ are defined in Eq.5.10, is a state-space realization of the DAE system.

A proof of the above proposition is obtained along the same lines as for the disturbance-free case in Proposition 3.2, by observing that for initial conditions $x(0) \in$

$\mathcal{M}$, and the solution for z in Eq.5.10 which is well-defined for any u and d, the variables $x(t)$ evolve in the constrained state space $\mathcal{M}$, irrespectively of the manipulated inputs u and the disturbances $d(t)$. Furthermore, similar to the result in Proposition 3.3, a state-space realization of dimension $\kappa = n - \sum_{i=1}^{s}(p - p_i)$ can be obtained in transformed coordinates, by using the linearly independent constraints $\mathbf{k}(x) = 0$ as a part of a nonlinear coordinate change.

For the DAE system in Eq.5.1, we will address the design of a dynamic feedforward/static state feedback controller of the form:

$$u = p(x) + q(x)v + Q(x, d, d^{(1)}, \ldots) \qquad (5.14)$$

to completely eliminate the effects of the measured disturbances d on the outputs y and induce a desired closed-loop input/output behavior between the outputs y and the reference inputs v, where $p(x)$ is a smooth vector field, $q(x)$ is a nonsingular matrix, and Q is a smooth function of the differential variables x, the disturbances d and their derivatives $d^{(1)}, \ldots$, which is nonsingular at nominal conditions $(x_0, 0, 0, \ldots)$ and well-defined for all smooth disturbances d.

Remark 5.1: Note that control law for the manipulated inputs u in Eq.5.14 may explicitly involve the disturbances d. However, due to the fact that the underlying algebraic constraints in x in a regular DAE system of Eq.5.1 do not involve the manipulated inputs u, the control law for u in Eq.5.14 does not introduce the disturbances d in the constraints in the closed-loop system. Thus, the closed-loop DAE system under a control law of the form in Eq.5.14 also has a well-defined state-space realization, irrespectively of the time-varying disturbances d.

With the objective of designing a feedforward/feedback controller of the form in Eq.5.14 for the DAE system in Eq.5.1, we will define the notions of relative orders of the controlled outputs y_i with respect to the manipulated inputs u and the disturbance inputs d. More specifically, we define the relative order r_i of the output y_i with respect to the manipulated input vector u, on the basis of the state-space realization in Eq.5.12, as the minimum integer such that:

$$[L_{\bar{g}_1}L_{\bar{f}}^{r_i-1}h_i(x) \ \ L_{\bar{g}_2}L_{\bar{f}}^{r_i-1}h_i(x) \ \ \cdots \ \ L_{\bar{g}_m}L_{\bar{f}}^{r_i-1}h_i(x)] \neq [0 \ \ 0 \ \cdots \ 0] \qquad (5.15)$$

for $x \in X$, where $X \subseteq \mathcal{M}$ is an open set containing the nominal equilibrium point of interest. If no such integer exists, then $r_i = \infty$. Similarly, the relative order ρ_i of the controlled output y_i with respect to the measured disturbance input vector d will be defined as the minimum integer such that:

$$[L_{\bar{\alpha}_1}L_{\bar{f}}^{\rho_i-1}h_i(x) \ \ L_{\bar{\alpha}_2}L_{\bar{f}}^{\rho_i-1}h_i(x) \ \ \cdots \ \ L_{\bar{\alpha}_\nu}L_{\bar{f}}^{\rho_i-1}h_i(x)] \neq [0 \ \ 0 \ \cdots \ 0] \qquad (5.16)$$

for $x \in X$. Again if no such integer exists, then $\rho_i = \infty$. In the above equations, $\bar{g}_i$ and $\bar{\alpha}_i$ denote the ith column vector of the respective matrices. It is assumed that a finite relative order r_i exists for each output y_i, since it is necessary for output controllability, and the $m \times m$ characteristic matrix:

$$C(x) = \begin{bmatrix} L_{\bar{g}_1} L_{\bar{f}}^{r_1-1} h_1(x) & \cdots & L_{\bar{g}_m} L_{\bar{f}}^{r_1-1} h_1(x) \\ \vdots & & \vdots \\ L_{\bar{g}_1} L_{\bar{f}}^{r_m-1} h_m(x) & \cdots & L_{\bar{g}_m} L_{\bar{f}}^{r_m-1} h_m(x) \end{bmatrix} \quad (5.17)$$

is nonsingular on X.

For the DAE system of Eq.1.9, with a state-space realization in Eq.5.12, relative orders r_i and ρ_i, and a nonsingular characteristic matrix $C(x)$, we now address the design of a feedforward/state feedback controller of the form in Eq.5.14 that completely eliminates the effects of the disturbances d on the controlled outputs y and induces an input/output response:

$$y + \sum_{i=1}^{m} \sum_{j=1}^{r_i} \gamma_{ij} \frac{d^j y_i}{dt^j} = y_{sp} \quad (5.18)$$

in the closed-loop system subject to the underlying constraints in Eq.5.8. Here $y_{sp} = [y_{1sp} \cdots y_{msp}]^T$ is the output setpoint vector and $\gamma_{ij} = [\gamma_{ij}^1 \cdots \gamma_{ij}^m]^T$ are vectors of adjustable parameters. The feedforward/state feedback controller synthesis problem for the DAE system of Eq.1.9 will be solved on the basis of the state-space realization of Eq.5.12, following the approach of [38] for nonlinear ODE systems. Theorem 5.1 that follows, states the controller synthesis result.

Theorem 5.1: *Consider a DAE system in Eq.5.1 with an equivalent state-space realization in Eq.5.12, for which $\det C(x) \neq 0$, $\forall x \in X$. Then, the following conditions:*

$$L_{\bar{g}_j} \phi_{il}(x, d(t)) \equiv 0, \quad i, j = 1, \ldots, m; \; l = 0, 1, \ldots, r_i - \rho_i - 1 \quad (5.19)$$

where:

$$\phi_{il}(x, d(t)) = \sum_{\mu=0}^{l} L_{\bar{f}}^{l-\mu} \left(\sum_{j=1}^{\nu} d_j(t) L_{\bar{\alpha}_j} + \frac{\partial}{\partial t} \right) \left(L_{\bar{f}} + \sum_{j=1}^{\nu} d_j(t) L_{\bar{\alpha}_j} + \frac{\partial}{\partial t} \right)^{\mu} L_{\bar{f}}^{\rho_i-1} h_i(x)$$

are necessary and sufficient for the existence of a dynamic feedforward/static state feedback law of the form in Eq.5.14 that induces the input/output behavior of Eq.5.18. If these conditions are satisfied, then the requisite control law takes the form:

$$u = \{[\gamma_{1r_1} \cdots \gamma_{mr_m}] C(x)\}^{-1} \left\{ y_{sp} - y - \sum_{i=1}^{m} \sum_{k=1}^{r_i} \gamma_{ik} L_{\bar{f}}^k h_i(x) \right.$$
$$\left. - \sum_{i=1}^{m} \sum_{k=\rho_i}^{r_i} \gamma_{ik} \phi_{i(k-\rho_i)} \left(x, d, d^{(1)}, \ldots, d^{(k-\rho_i)} \right) \right\} \quad (5.20)$$

Proof: A proof of the conditions in Eq.5.19 can be derived on the basis of the state-space realization (Eq.5.12) of the DAE system in Eq.5.1, following an approach similar to [38]. For systems satisfying these conditions, the result on the necessary control law in Eq.5.20 can be established on the basis of a state-space realization of the closed-loop DAE system. Clearly, for a regular DAE system in Eq.5.1 satisfying the condition $\nu_{q,1} = p_q$ for all $q \geq 2$, this condition is also satisfied for the closed-loop DAE system obtained under the control law of Eq.5.20. Thus, the algorithm for the closed-loop system also proceeds exactly the same as the open-loop system, identifying the same constraints in x and solution for z, to yield the state-space realization in Eq.5.12 with u as in Eq.5.20. Thereafter, the proof follows similar to [38], evaluating the expressions for the output derivatives and verifying the requested input/output response in Eq.5.18. □

The closed-loop stability of the DAE system in Eq.5.1 under the controller in Eq.5.20 is ensured, if the system is minimum-phase, i.e., the zero dynamics of Eq.5.12 is locally asymptotically stable, and the adjustable parameters γ_{ij}^k are chosen such that the input/output response requested in Eq.5.18 is BIBO (bounded-input bounded-output) stable. It should be mentioned that depending on the relative orders r_i and ρ_i, the control law in Eq.5.20 may involve the derivatives of the disturbances d. In the practical implementation of such control laws, the disturbance derivatives will be approximated through suitable lead-lag filters.

5.5 Feedforward/State Feedback Regularization and Control of Nonregular Systems

For a nonregular DAE system of Eq.5.1 with a finite index ν_d and satisfying the condition $\nu_{q,2} = m_q$ for all $q \geq 2$, the algorithm will converge in a finite number of iterations s ($s < \nu_d - 1$) with $p_1 \leq p_2 \leq \cdots \leq p_{s+1} < p$ and $m_2 \leq m_3 \leq \cdots \leq m_{s+1} = p$, ($m_i \geq p_i, \forall i > 1$). The algorithm identifies the following linearly independent constraints in x that do not involve the manipulated inputs u and the disturbances d:

$$\mathbf{k}(x) = \begin{bmatrix} \mathbf{k}^1(x) \\ \vdots \\ \mathbf{k}^s(x) \end{bmatrix} = 0 \tag{5.21}$$

Moreover, the algorithm yields the algebraic equations:

$$0 = \begin{bmatrix} \overline{k}^s(x) \\ \widehat{k}^s(x) \\ \widetilde{k}^{s+1}(x) \end{bmatrix} + \begin{bmatrix} \overline{l}^s(x) \\ 0 \\ \widetilde{l}^{s+1}(x) \end{bmatrix} z + \begin{bmatrix} \overline{c}^s(x) \\ \widehat{c}^s(x) \\ \widetilde{c}^{s+1}(x) \end{bmatrix} u + \begin{bmatrix} \overline{\beta}^s(x) \\ \widehat{\beta}^s(x) \\ \widetilde{\beta}^{s+1}(x) \end{bmatrix} d \tag{5.22}$$

81

where the matrices:

$$L^{s+1}(x) = \begin{bmatrix} \bar{l}^s(x) \\ \tilde{l}^{s+1}(x) \end{bmatrix}, \quad L^{s+1,e}(x) = \begin{bmatrix} \bar{l}^s(x) & \bar{c}^s(x) \\ 0 & \hat{c}^s(x) \\ \tilde{l}^{s+1}(x) & \tilde{c}^{s+1}(x) \end{bmatrix}$$

have ranks $p_{s+1} < p$ and $m_{s+1} = p$, respectively. Premultiplying the above algebraic equations with a $p \times p$ nonsingular matrix $E^{s+1}(x)$, the following final set of algebraic equations are obtained:

$$0 = \begin{bmatrix} \bar{k}(x) \\ \hat{k}(x) \end{bmatrix} + \begin{bmatrix} \bar{l}(x) \\ 0 \end{bmatrix} z + \begin{bmatrix} \bar{c}(x) \\ \hat{c}(x) \end{bmatrix} u + \begin{bmatrix} \bar{\beta}(x) \\ \hat{\beta}(x) \end{bmatrix} d \qquad (5.23)$$

where $\bar{k}(x), \hat{k}(x)$ are vector fields of dimensions $p_{s+1}, (p - p_{s+1})$, $\bar{c}(x), \hat{c}(x), \bar{\beta}(x), \hat{\beta}(x)$ are matrices of dimensions $p_{s+1} \times m$, $(p - p_{s+1}) \times m$, $p_{s+1} \times \nu$ and $(p - p_{s+1}) \times \nu$, respectively, and the $p_{s+1} \times p$, $p \times (p + m)$ matrices:

$$\bar{l}(x), \quad \begin{bmatrix} \bar{l}(x) & \bar{c}(x) \\ 0 & \hat{c}(x) \end{bmatrix}$$

have full row rank. Thus, the algorithm yields the following new DAE system:

$$\dot{x} = f(x) + b(x)z + g(x)u + \alpha(x)d$$
$$0 = \begin{bmatrix} \bar{k}(x) \\ \hat{k}(x) \end{bmatrix} + \begin{bmatrix} \bar{l}(x) \\ 0 \end{bmatrix} z + \begin{bmatrix} \bar{c}(x) \\ \hat{c}(x) \end{bmatrix} u + \begin{bmatrix} \bar{\beta}(x) \\ \hat{\beta}(x) \end{bmatrix} d$$
$$y_i = h_i(x), \quad i = 1, \ldots, m \qquad (5.24)$$

where $x \in \mathcal{X} : \mathbf{k}(x) = 0$. The DAE system in Eq.5.24 is equivalent to the DAE system in Eq.5.1, i.e., for consistent initial conditions $x(0)$ such that $\mathbf{k}(x(0)) = 0$, and sufficiently smooth inputs $u(t), d(t)$, both systems have the same solution $(x(t), z(t))$. The algebraic equations in the new DAE system (Eq.5.24) explicitly include the constraints in x, $0 = \hat{k}(x) + \hat{c}(x)u + \hat{\beta}(x)d$, that involve the manipulated inputs u and possibly the disturbance inputs d.

Following the result on the design of dynamic feedback regularizing compensator in Section 4.4, we will address the design of a feedforward/dynamic state feedback compensator of the form:

$$\dot{w} = \mathcal{W}(x, w, v, d)$$
$$u = \mathcal{F}(x, w, v, d) \qquad (5.25)$$

where $v \in \mathbb{R}^m$ is the vector of new (manipulated) inputs and $w \in \mathbb{R}^{n_c}$ is the vector of compensator states, to eliminate the disturbances d from the algebraic constraints and obtain a modified DAE system that is regular. The following lemma states the result on the class of nonregular DAE systems in Eq.5.1 for which such a feedforward/feedback regularization is possible.

Lemma 5.2: *Consider a nonregular DAE system in Eq.5.1 with a finite index ν_d, for which the matrices $L^q(x)$, $L^{q,e}(x)$ and $L^{q,eII}(x)$ in Eq.5.4 have ranks p_q, $m_q \geq p_q$ and $\nu_{q,2} \geq m_q$ in iteration $q \geq 2$ of the algorithm. Then, the disturbances d can be eliminated from the algebraic constraints in the differential variables through a feedforward/dynamic state feedback compensator of the form in Eq.5.25, to obtain a modified system that is regular, if and only if, the condition $\nu_{q,2} = m_q$ holds for each iteration $q \geq 2$ of the algorithm.*

The proof of necessity is quite straightforward and follows from the fact that if $\nu_{q,2} > m_q$ in any iteration, then there are constraints in x which involve the disturbances d but not the manipulated inputs u. Clearly, the disturbances can not be eliminated from these constraints through any feedforward/feedback compensator of the form in Eq.5.25. The sufficiency follows from the construction of a compensator. The design of the requisite feedforward/feedback compensator involves (i) an x-dependent input transformation $u = M(x)[\bar{u}_1^T \;\; \bar{u}_2^T]^T$, where $\bar{u}_1 \in \mathbb{R}^{(p-p_{s+1})}$, $\bar{u}_2 \in \mathbb{R}^{m-(p-p_{s+1})}$ are the new inputs and $M(x)$ is a nonsingular $m \times m$ matrix chosen such that in the relation:

$$\begin{bmatrix} \bar{c}(x) \\ \hat{c}(x) \end{bmatrix} M(x) = \begin{bmatrix} \bar{c}_1(x) & \bar{c}_2(x) \\ \hat{c}_1(x) & 0 \end{bmatrix} \tag{5.26}$$

the $(p-p_{s+1}) \times (p-p_{s+1})$ matrix $\hat{c}_1(x)$ is nonsingular, thereby isolating the inputs $\bar{u}_1$ that appear in the algebraic constraints in a nonsingular fashion, and (ii) designing a feedforward/dynamic state feedback compensator for these inputs $\bar{u}_1$ such that the corresponding constraints in the modified system are independent of the new inputs v and the disturbances d, and one differentiation of these constraints yields a set of algebraic equations that can be solved for z. The objective in (ii) is achieved by modifying the constraints $0 = \hat{k}(x) + \hat{c}_1(x)\bar{u}_1 + \hat{\beta}(x)d$ in the DAE system of Eq.5.24 through a feedforward/feedback law for the inputs $\bar{u}_1$, to the constraints $Sx + w = 0$ such that:

$$\text{rank} \begin{bmatrix} \bar{l}(x) \\ Sb(x) \end{bmatrix} = p \tag{5.27}$$

where $S \in \mathbb{R}^{(p-p_{s+1}) \times n}$ is a constant matrix. The existence of a matrix S satisfying the above condition is ensured from the result of Lemma 4.1. Theorem 5.2 that follows, states the result on the feedforward/dynamic state feedback regularizing compensator.

Theorem 5.2: *Consider a nonregular DAE system in Eq.5.1, for which the algorithm converges in s iterations with $\nu_{q,2} = m_q$ for all $q \geq 2$, to yield the equivalent DAE system* *in Eq.5.24. Then, the feedforward/dynamic state feedback compensator:*

$$\dot{w} = v_1$$

$$u = M(x)\left[\begin{array}{c} (\widehat{c}_1(x))^{-1}(-\widehat{k}(x) - \widehat{\beta}(x)d + Sx + w) \\ v_2 \end{array}\right] \tag{5.28}$$

where $w, v_1 \in \mathbb{R}^{(p-p_s+1)}$, $v_2 \in \mathbb{R}^{m-(p-p_s+1)}$, and the matrices $M(x)$ and S are chosen as in Eq.5.26 and Eq.5.27, respectively, yields the modified DAE system:

$$\dot{x} = \widetilde{f}(x,w) + b(x)z + \overline{g}_2(x)v_2 + \widetilde{\alpha}(x)d$$

$$\dot{w} = v_1$$

$$0 = \left[\begin{array}{c} \widetilde{k}(x,w) \\ Sx + w \end{array}\right] + \left[\begin{array}{c} \widetilde{l}(x) \\ 0 \end{array}\right]z + \left[\begin{array}{cc} 0 & \overline{c}_2(x) \\ 0 & 0 \end{array}\right]\left[\begin{array}{c} v_1 \\ v_2 \end{array}\right] + \left[\begin{array}{c} \widetilde{\beta}(x) \\ 0 \end{array}\right]d$$

$$y_i = h_i(x), \quad i = 1,\ldots,m \tag{5.29}$$

where:

$$\widetilde{f}(x,w) = f(x) + \overline{g}_1(x)\gamma(x) + \overline{g}_1(x)(\widehat{c}_1(x))^{-1}w$$
$$\widetilde{\alpha}(x) = \alpha(x) - \overline{g}_1(x)(\widehat{c}_1(x))^{-1}\widehat{\beta}(x)$$
$$\widetilde{k}(x,w) = \overline{k}(x) + \overline{c}_1(x)\gamma(x) + \overline{c}_1(x)(\widehat{c}_1(x))^{-1}w$$
$$\widetilde{\beta}(x) = \overline{\beta}(x) - \overline{c}_1(x)(\widehat{c}_1(x))^{-1}\widehat{\beta}(x)$$
$$\gamma(x) = (\widehat{c}_1(x))^{-1}\left\{-\widehat{k}(x) + Sx\right\}$$
$$[\overline{g}_1(x) \quad \overline{g}_2(x)] = g(x)M(x)$$

and $x \in \mathcal{X}$: $\mathbf{k}(x) = 0$. The DAE system in Eq.5.29 with an extended vector of differential variables $\bar{x} = [x^T \ w^T]^T$ and a new vector of inputs $v = [v_1^T \ v_2^T]^T \in \mathbb{R}^m$ is solvable with a finite index and the underlying constraints in $\bar{x}$ are independent of the inputs v and the disturbances d.

Proof: The DAE system of Eq.5.24, under the proposed compensator of Eq.5.28, directly yields the extended DAE system of Eq.5.29, with the corresponding modified constraints in the differential variables $\bar{x}$, $Sx + w = 0$, which are independent of the new inputs v and the disturbances d. Differentiating these constraints once, in the resulting algebraic equations, the coefficient matrix for z is the same as in Eq.4.24, which is nonsingular. Thus, the DAE system of Eq.4.27 is solvable, with an index $\nu_d = 2$ and a unique smooth solution for z of the form:

$$z = R(x,w) + S_1(x)v_1 + S_2(x)v_2 + S_3(x)d \tag{5.30}$$

Moreover, the algebraic constraints in the differential variables $\bar{x}$, i.e., $\mathbf{k}(x) = 0$, $Sx + w = 0$, are independent of the new inputs v and the disturbances d, and they specify the constrained subspace:

$$\mathcal{M} = \left\{(x,w) \in \mathcal{X} \times \mathbb{R}^{(p-p_s+1)} \ : \ \begin{array}{c} \mathbf{k}(x) = 0 \\ Sx + w = 0 \end{array}\right\} \tag{5.31}$$

of dimension $\kappa = n - (p - p_1) - \sum_{i=2}^{s}(p - m_i)$, which can be easily verified to be invariant under any control law for the inputs v and any disturbance d. This completes the proof. □

For the feedforward/feedback regularized DAE system in Eq.5.29, the reconstruction of the algebraic variables z (Eq.5.30) and the specification of the constrained state space $\mathcal{M}$ (Eq.5.31), allow the derivation of a state-space realization, which is stated in the following proposition.

Proposition 5.2: *Consider a nonregular DAE system of Eq.5.1, for which the algorithm yields the equivalent DAE system of Eq.5.24 and the compensator of Theorem 5.2 yields the regular extended DAE system of Eq.5.29. Then, the dynamic system:*

$$\begin{aligned}
\dot{\bar{x}} &= \overline{f}(\bar{x}) + \overline{g}(\bar{x})v + \overline{\alpha}(\bar{x})d \\
y_i &= \overline{h}_i(\bar{x}), \ i = 1, \ldots, m
\end{aligned} \tag{5.32}$$

is a state-space realization of the regularized DAE system in Eq.5.29, where $\bar{x} = [x^T \ w^T]^T \in \mathcal{M}$ is the extended state vector, $v \in \mathbb{R}^m$ is the new input vector defined in Theorem 5.2, and

$$\overline{f}(\bar{x}) = \begin{bmatrix} \widetilde{f}(x, w) + b(x)R(x, w) \\ 0 \end{bmatrix}, \quad \overline{\alpha}(\bar{x}) = \begin{bmatrix} \widetilde{\alpha}(x) + b(x)S_3(x) \\ 0 \end{bmatrix}$$

$$\overline{g}(\bar{x}) = \begin{bmatrix} b(x)S_1(x) & \overline{g}_2(x) + b(x)S_2(x) \\ I_{p-p_s+1} & 0 \end{bmatrix}, \quad \overline{h}_i(\bar{x}) = h_i(x)$$

The state-space realization of Eq.5.32 of the regularized DAE system (Eq.5.29) can be used to introduce the notions of relative orders r_i and ρ_i of the output y_i with respect to the input vector v and disturbance vector d, respectively. Then, following the approach in the previous section for regular systems, a dynamic feedforward/static state feedback control law can be derived for the inputs v to eliminate the effects of the disturbances d on the controlled outputs y_i and enforce a linear input/output behavior of Eq.5.18 with closed-loop stability. Depending on the relative orders ρ_i and r_i, the control law for the inputs v may explicitly involve the disturbances d. However, similar to the regular case (see Remark 5.1), the control law will not introduce these disturbances in the algebraic constraints in the closed-loop extended DAE system, i.e., the closed-loop DAE system will have a well-defined state-space realization irrespectively of the time-varying disturbances. The dynamic feedforward/static feedback control law for the inputs v and the feedforward/dynamic state feedback compensator of Theorem 5.2, will comprise the overall dynamic feedforward/state feedback controller for the DAE system in Eq.5.1.

5.6 Conclusions

In this chapter, we addressed the control of nonlinear high-index DAE systems, with external time-varying disturbance inputs that are modeled explicitly. The presence of arbitrarily time-varying disturbances may lead to problems in the existence of a well-defined index and state-space realizations, specifically if these disturbances appear in the underlying algebraic constraints in the differential variables. Motivated by this, we identified and focused on the class of (i) regular DAE systems for which the algebraic constraints are independent of the disturbances, and thus, a state-space realization exists irrespectively of the disturbances, and (ii) nonregular DAE systems for which the measured disturbances appear only in those algebraic constraints that also involve the manipulated inputs. For the latter class of systems, the disturbances can be eliminated from the algebraic constraints through a feedforward/feedback regularizing compensator, to obtain a regular DAE system for which a well-defined state-space realization exists despite the presence of the disturbances. The derived state-space realizations were then used to address the synthesis of feedforward/state feedback controllers that induce a desired closed-loop response where the effect of the measured disturbances on the controlled outputs is eliminated.

Notation

Roman letters

b, g, l, c	=	matrices in the DAE system
C	=	characteristic matrix
E^i	=	$p \times p$ nonsingular matrices
d	=	vector of disturbance inputs
f, k	=	vector fields in the DAE system
$\overline{f}$	=	vector field in state-space realization of dimension n
$\overline{g}$	=	matrix in state-space realization of dimension n
$\overline{k}^i, \mathbf{k}^i, \widetilde{k}^i$	=	vector fields in the algorithm
$\overline{l}^i, \widetilde{l}^i, \overline{c}^i, \widetilde{c}^i$	=	matrices in the algorithm
h_i	=	scalar functions in the DAE system
$\mathcal{M}$	=	constrained manifold where differential variables evolve
m	=	number of manipulated inputs and controlled outputs
n	=	number of differential variables
p	=	number of algebraic variables
p_i, m_i	=	ranks of matrices in algorithm
r_i	=	relative order of output y_i with respect to u
s	=	number of iterations for convergence of algorithm
u	=	manipulated input vector
v	=	external input vector
x	=	vector of differential variables
y_i	=	output variable
y_{sp}	=	vector of output setpoints
z	=	vector of algebraic variables

Greek letters

$\alpha, \overline{\alpha}$	=	matrices in DAE system and state-space realization
$\overline{\beta}^i, \widehat{\beta}^i, \widetilde{\beta}^i$	=	matrices in algorithm
γ_{ij}	=	vectors of adjustable controller parameters
ρ_i	=	relative order of output y_i with respect to d
ν	=	number of disturbance inputs
$\nu_{i,1}, \nu_{i,2}$	=	ranks of matrices in algorithm
ν_d	=	index of a DAE system

Math symbols

$$\mathbb{R} \quad = \quad \text{real line}$$

$$\mathbb{R}^i \quad = \quad i\text{-dimensional Euclidean space}$$

$$[\,\cdot\,]^T \quad = \quad \text{transpose of a vector/matrix}$$

$L_f\alpha(x) \quad = \quad$ Lie derivative of a scalar function $\alpha(x)$ with respect to vector field $f(x)$, defined as $L_f\alpha(x) = [\dfrac{\partial\alpha}{\partial x_1} \quad \cdots \quad \dfrac{\partial\alpha}{\partial x_n}]f(x)$

$L_f^j\alpha(x) \quad = \quad$ high-order Lie derivative defined as $L_f^j\alpha(x) = L_f(L_f^{j-1}\alpha(x))$

88

6. DAEs and Singularly Perturbed Systems

6.1 Introduction

Singularly perturbed (ODE) systems arise as models of a wide variety of chemical processes [7, 42], biochemical systems [3, 59], electrical circuits and power systems [70, 114], etc., whose dynamic behavior exhibits time-scale multiplicity. There is, as one may expect, an inherent connection between DAEs and singularly perturbed systems. More specifically, singularly perturbed systems are characterized by the presence of a small parameter ϵ, known as the singular perturbation parameter, which when approximated to zero, leads to an abrupt reduction in the order of the system dynamics; these reduced order dynamics are in general described by a DAE system.

There is a vast literature on the application of singular perturbation theory for the analysis and control of systems with multiple time scales, in particular two time scales (see, e.g., [71, 73, 74, 109]). For nonlinear two-time-scale systems, it is well-established that standard inversion-based controllers, designed without explicitly addressing the time-scale multiplicity, are often ill-conditioned and may lead to instability [74]. Singular perturbation theory allows addressing the control of two-time-scale systems through a systematic decomposition of the system dynamics in different time scales.

The vast majority of available research has focused on two-time-scale systems modeled in the so-called "standard" singularly perturbed form, where the singular perturbation parameter ϵ multiplies the time-derivative of the "fast" state variables. For such systems, the stability analysis and the design of well-conditioned controllers is addressed on the basis of reduced-order representations of slow and fast dynamics in separate slow and fast time scales (see, e.g., [27–29, 71, 72, 74] and the references therein). More specifically, a description for the slow dynamics is obtained in the limit $\epsilon \to 0$, when the dynamics of the fast variables become instantaneous and the corresponding differential equations reduce to a set of quasi-steady-state algebraic equations. Similarly, a reduced-order description for the fast dynamics is obtained in the limit $\epsilon \to 0$ in a "stretched" or fast time scale, when the slow variables are constant. The controller synthesis is then addressed through a combination of separate components designed to (i) stabilize the fast dynamics, if they are unstable, and (ii) achieve desired closed-loop performance objectives on the basis of the slow subsystem, which essentially governs the input/output behavior of the two-time-scale system. Such a controller design approach involving a combination of slow and fast controllers for the respective time scales is known as composite control [23, 24, 127].

A key requirement for the application of the available results is that the two-time-scale process must be modeled by an ODE system in the standard singularly perturbed form. However, this is often a nontrivial task in itself. In some chemical processes, e.g., catalytic reactors [21] and fluidized catalytic cracking [42], there is an *a priori* knowledge of the variables with slow and fast dynamics; they are typically associated with large and small holdups or heat capacities. This allows modeling such processes directly in the standard singularly perturbed form, through an appropriate definition of ϵ, e.g., the small holdup or the reciprocal of the large holdup. On the other hand, for the fast-rate processes discussed in Chapter 2, the fast dynamics corresponding to the fast reactions, fast mass transfer, etc., can not be associated with distinct state variables, and the detailed rate-based models are *not* in the standard singularly perturbed form. The derivation of a standard singularly perturbed representation for such systems requires, in general, a nonlinear coordinate change.

Early research on nonlinear ODE systems with a small parameter ϵ, not necessarily in the standard singularly perturbed form, focused on studying their geometric properties to obtain a coordinate-free characterization of time-scale multiplicity [44]. This characterization was subsequently used to derive necessary and sufficient geometric conditions for the existence of an ϵ-independent change of coordinates that yields a standard singularly perturbed representation of autonomous systems [113, 114] and control systems with manipulated inputs [100]. On the other hand, for a class of singularly perturbed systems in nonstandard form, the connection with a corresponding class of high-index DAE systems was used to propose an ϵ-dependent coordinate change that is singular at $\epsilon = 0$ and yields a standard singularly perturbed representation [78].

In this chapter, we briefly review singularly perturbed systems in standard form and then address the analysis and control of a class of nonstandard singularly perturbed systems that arise as rate-based models of the fast-rate chemical processes discussed in Chapter 2. These rate-based ODE models are characterized by the presence of large parameters of the form $1/\epsilon$ in the rate expressions, e.g., large mass transfer/reaction rate coefficients. We perform a singular perturbation analysis of the rate-based models, to identify regions in state space where the fast-rate process exhibits a two-time-scale behavior, and obtain a standard singularly perturbed representation of the model in this region. Initially, we focus on the construction of an ϵ-independent coordinate change that yields such a representation. This is possible only for a specific class of systems for which certain necessary and sufficient conditions are satisfied and the two-time-scale property holds in the whole state space region [82, 83]. For another class of systems where these conditions are violated, the two-time-scale property holds only in a state space region that shrinks with ϵ, and correspondingly, the requisite coordinate change must depend on ϵ in a singular fashion. Finally, we outline directions for further work for more general classes of singularly perturbed systems arising as dynamic models of process networks, which may possibly have more than two time scales.

The derivation of the standard singularly perturbed representation of the rate-based ODE models allows the application of available results on stability analysis and control for two-time-scale systems. In particular, the fast dynamics associated with the fast reactions, mass transfer, etc., are typically stable. This implies that these stable fast dynamics can be ignored, and the synthesis of well-conditioned controllers can be addressed on the basis of the representation of the slow dynamics. In the equilibrium-based DAE models of Chapter 2, the assumptions of phase, reaction, pressure equilibrium, etc., precisely amount to ignoring the fast dynamics, i.e., the high-index DAE models describe the slow dynamics of the fast-rate processes. Thus, controllers designed on the basis of the equilibrium-based high-index DAE models do not suffer from the aforementioned problems that arise in controllers designed on the basis of the rate-based ODE/index-one DAE models.

6.2 Standard vs. Nonstandard Singularly Perturbed Systems

The majority of research on the analysis and control of two-time-scale systems has focused on systems with the following general description:

$$\dot{\zeta} = F(\zeta, \eta, u, \epsilon)$$
$$\epsilon\dot{\eta} = G(\zeta, \eta, u, \epsilon)$$
$$y_i = h_i(\zeta, \eta), \ i = 1, \ldots, m \tag{6.1}$$

where $\zeta \in \mathcal{Z} \subset \mathbb{R}^{n_1}$, $\eta \in \mathcal{Y} \subset \mathbb{R}^{n_2}$ are the state variables with $\mathcal{Z}, \mathcal{Y}$ being open and connected sets, $u \in \mathbb{R}^m$ is the vector of manipulated inputs, y_i's are the outputs to be controlled, F, G are smooth vector fields of dimensions n_1 and n_2, respectively, and h_i are smooth scalar functions. The system description in Eq.6.1 is characterized by the presence of the small positive parameter ϵ, known as the singular perturbation parameter, which multiplies the time-derivative of the state variables η. In the limiting case when $\epsilon \to 0$, the differential equations for η reduce to a set of algebraic equations $G(\zeta, \eta, u, 0) = 0$, i.e., the order of the dynamic system reduces abruptly from $n_1 + n_2$ to n_1. Throughout this chapter, we will use the standard order of magnitude notation $O(\epsilon)$, where $\delta(\epsilon) = O(\epsilon)$ if there exist positive constants k and c such that: $|\delta(\epsilon)| \leq k|\epsilon|$, $\forall |\epsilon| < c$.

The above singularly perturbed system is said to be in the standard form if $G(\zeta, \eta, u, 0) = 0$ has $k \geq 1$ isolated real roots $\eta_i = \alpha_i(\zeta, u)$, $i = 1, \ldots, k$, i.e., the Jacobian $(\partial G(\zeta, \eta, u, 0)/\partial \eta)$ is nonsingular. It should be mentioned that it is quite possible that the original system may not be in standard form, but could be transformed into standard form through a preliminary feedback of the form $u = \tilde{u} + \mathcal{F}(\eta)$ under appropriate conditions (for more details, see, e.g., [69]). Owing to the presence of the small parameter ϵ that multiplies the time-derivative of η, the system of Eq.6.1

is characterized by an explicit time scale separation with the states η being the fast ones and the states ζ being the slow ones. A time-scale decomposition of the above singularly perturbed system yields reduced-order representations for the slow and fast subsystems.

More specifically, the following description of the slow dynamics in the slow time scale t is obtained in the limit $\epsilon \to 0$, when the fast dynamics for the states η become instantaneous and the corresponding differential equations reduce to algebraic equations:

$$\dot{\zeta} = F(\zeta, \eta, u, 0)$$
$$0 = G(\zeta, \eta, u, 0) \tag{6.2}$$

Clearly the above system is a DAE system with index $\nu_d = 1$ since the algebraic equations $0 = G(\zeta, \eta, u, 0)$ can be solved for the quasi-steady-state solution $\eta = \alpha(\zeta, u)$. Substituting the solution for η in the differential equations for ζ, the following reduced-order representation of the slow subsystem is obtained:

$$\dot{\zeta} = F(\zeta, \alpha(\zeta, u), u, 0) \tag{6.3}$$

The slow subsystem is also referred to as a *reduced* or *quasi-steady-state* subsystem.

A representation of the fast subsystem is similarly obtained in the limit $\epsilon \to 0$ in a "stretched" fast time scale $\tau = t/\epsilon$. In this fast time scale, τ, the system of Eq.6.1 has the following description:

$$\frac{d\zeta}{d\tau} = \epsilon F(\zeta, \eta, u, \epsilon)$$
$$\frac{d\eta}{d\tau} = G(\zeta, \eta, u, \epsilon) \tag{6.4}$$

In the limit $\epsilon \to 0$, the dynamics of the slow variables ζ become negligible and the following reduced-order representation is obtained for the fast subsystem:

$$\frac{d\eta}{d\tau} = G(\zeta, \eta, u, 0) \tag{6.5}$$

where the slow variables ζ are frozen at their initial value $\zeta(0)$ and treated as constant parameters. The fast subsystem is also referred to as the *boundary-layer* subsystem owing to the fact that if the fast dynamics of the variables η are stable, then these variables converge from the initial condition $\eta(0)$ to the quasi-steady-state value $\alpha(\zeta, u)$ in the fast time scale $\tau = O(\epsilon)$. Often, the fast subsystem representation is expressed in terms of the deviation variables $\eta_f = \eta - \alpha(\zeta, u)$:

$$\frac{d\eta_f}{d\tau} = G(\zeta, \alpha(\zeta, u) + \eta_f, u, 0) \tag{6.6}$$

such that $\eta_f = 0$ is the quasi-steady-state solution.

92

In a general nonlinear system, there may be several distinct solutions $\eta_i = \alpha_i(\zeta, u)$, $i = 1, \ldots, k$ of $G(\zeta, \eta, u, 0) = 0$. In such a case, one focuses on a particular solution and the corresponding representation for the slow subsystem in an appropriate neighborhood. The specific choice of a particular quasi-steady-state solution depends on the initial condition $\zeta(0), \eta(0)$; the solution $\eta(\tau)$ of the fast subsystem will stabilize at a corresponding quasi-steady-state $\eta_i = \alpha_i(\zeta(0), u)$.

Singular perturbation theory allows inferring the asymptotic properties of the system in Eq.6.1 with a small, but nonzero ϵ, from the analysis of the asymptotic properties of the lower-order slow (Eq.6.3) and fast (Eq.6.5) subsystems obtained in the limit $\epsilon = 0$ in the respective time scales [73]. A key result that is the cornerstone of the extensive literature on control of singularly perturbed systems is the result of Tikhonov [131], which expresses the relationship between the true solution $\zeta(t), \eta(t)$ of the system in Eq.6.1 with initial conditions $\zeta(0), \eta(0)$, and the solutions $\bar{\zeta}(t)$ and $\bar{\eta}_f(\tau)$ of the slow (Eq.6.3) and fast (Eq.6.6) subsystems, respectively, with the corresponding initial conditions $\zeta(0)$ and $\eta(0) - \alpha(\zeta(0), u)$. More specifically, the theorem states that if the fast subsystem is locally exponentially stable, and the slow subsystem has a unique solution on the time-interval $t \in [0, t_1]$, then for a sufficiently small $\epsilon < \epsilon^*$, $\zeta(t) - \bar{\zeta}(t) = 0(\epsilon)$ and $\eta(t) - \alpha(\bar{\zeta}(t), u) - \bar{\eta}_f(t/\epsilon) = 0(\epsilon)$ holds uniformly for $t \in [0, t_1]$. While the basic result holds only on bounded time intervals of $0(1)$, a similar result holds on the infinite time-interval under the additional assumption that the slow subsystem is also locally exponentially stable. For a detailed exposition on these results see, e.g., [70]. The result from Tikhonov's theorem thus enables the analysis and controller design for the system in Eq.6.1 on the basis of the low-order slow and fast subsystems, i.e., considering only the behavior of the system in the limit $\epsilon \to 0$ in separate time scales, thereby avoiding the singularity with respect to ϵ.

Remark 6.1: The result of Tikhonov's theorem yields an $0(\epsilon)$ approximation for the solution of the standard form singularly perturbed system in Eq.6.1. This is typically sufficient for addressing the controller synthesis on the basis of the slow subsystem, which ensures that within an $0(\epsilon)$ approximation, a desired tracking performance is enforced in the overall two-time-scale system. However, if desired, a solution with higher-order accuracy can also be obtained. More specifically, under appropriate stability conditions on the fast dynamics, the solution of Eq.6.1 has the general form (see, e.g., [109]):

$$\zeta(t, \epsilon) = \zeta^o(t, \epsilon) + \epsilon \zeta^i(\tau, \epsilon)$$
$$\eta(t, \epsilon) = \eta^o(t, \epsilon) + \eta^i(\tau, \epsilon) \qquad (6.7)$$

where ζ^i, η^i is the "inner" solution in the fast boundary layer and ζ^o, η^o is the "outer" solution after the boundary layer, with the following Taylor series expansion:

$$\zeta^i(\tau, \epsilon) = \sum_{j=0}^{\infty} \zeta^i_j(\tau)\epsilon^j, \quad \eta^i(\tau, \epsilon) = \sum_{j=0}^{\infty} \eta^i_j(\tau)\epsilon^j$$

$$\zeta^o(t, \epsilon) = \sum_{j=0}^{\infty} \zeta^o_j(t)\epsilon^j, \quad \eta^o(t, \epsilon) = \sum_{j=0}^{\infty} \eta^o_j(t)\epsilon^j$$

(6.8)

In the above series expansion, $\zeta^o_j, \eta^o_j, \zeta^i_j, \eta^i_j$ are smooth ϵ-independent coefficients where $\zeta^i_j(\tau)$ and $\eta^i_j(\tau)$ decay exponentially to zero in the boundary layer as $\tau \to \infty$. Thus, after the initial boundary layer, the solution of Eq.6.1 is given by the outer solution satisfying:

$$\dot{\zeta}^o = F(\zeta^o, \eta^o, t, \epsilon)$$
$$\dot{\eta}^o = G(\zeta^o, \eta^o, t, \epsilon)$$

(6.9)

Substituting the series expression for ζ^o and η^o and equating the coefficients of like powers of ϵ yields the governing equations for ζ^o_j and η^o_j. In particular, ζ^o_0, η^o_0 comprise a solution with an error of $O(\epsilon)$ and are governed by the index-one DAE system of Eq.6.2. However, higher-order solutions with an accuracy of $O(\epsilon^k)$, $k \geq 2$ require solving for the coefficients ζ^o_j, η^o_j, $j = 0, \ldots, k - 1$, which are governed by a DAE system with an increasingly higher index k [57].

Clearly, the assumption of a nonsingular Jacobian $(\partial G(\zeta, \eta, u, 0)/\partial \eta)$ is critical in the derivation of the slow and fast subsystems and the subsequent analysis. Whenever this assumption is violated, the system in Eq.6.1 is said to be nonstandard or "singular" singularly perturbed system [13, 118, 119], and the available results for standard singularly perturbed systems are not directly applicable. Moreover, unlike systems in the standard form, for systems in nonstandard form, the DAE system of Eq.6.2 obtained in the limit $\epsilon \to 0$ has a high index $\nu_d > 1$.

For singularly perturbed systems in nonstandard form, the separation of slow and fast variables is not explicit. Thus, a key problem for such systems is to obtain a standard form representation, if possible, through a suitable change of coordinates, thereby explicitly identifying the slow and fast variables in the new coordinates. In [100] systems of the general form:

$$\frac{dx}{d\tau} = f(x, u, \epsilon)$$

(6.10)

where $x \in \mathbb{R}^n$, were considered in the fast time scale τ and their two-time-scale properties were studied. For these systems, necessary and sufficient geometric conditions were derived for the existence of an ϵ-independent coordinate change of the form: $\zeta = \phi_s(x)$, $\eta = \phi_f(x)$ that transforms them into an equivalent system in the standard singularly perturbed form. Owing to the generality of the form in Eq.6.10, these

94

goometric conditions: (*i*) the existence of a possibly control-dependent equilibrium manifold $E^u = \{x \in \mathbb{R}^n : f(x, u, 0) = 0\}$, (*ii*) the existence of a family of conservation manifolds $C = \{x \in \mathbb{R}^n : \phi_s(x) = c\}$ where $c \in \mathbb{R}^{n_1}$, $n_1 < n$ and (*iii*) the transversality of the equilibrium and conservation manifolds, were expressed in an abstract coordinate-free setting.

In what follows, we will focus on a class of systems in nonstandard singularly perturbed form, arising as detailed dynamic models of the fast-rate chemical processes discussed in Chapter 2. The detailed rate-based models of these processes with fast mass transfer, fast reactions, etc., are characterized by the presence of large parameters, e.g., large mass transfer, reaction rate coefficients, in the explicit rate expressions. Defining ϵ as the reciprocal of such a representative large parameter, the detailed dynamic models of these fast-rate processes are given by a system with the following general description:

$$\dot{x} = f(x) + g(x)u + \frac{1}{\epsilon}b(x)k(x)$$
$$y_i = h_i(x), \quad i = 1, \ldots, m \qquad (6.11)$$

where $x \in \mathcal{X} \subset \mathbb{R}^n$ is the vector of state variables, $f(x)$, $k(x)$ are smooth vector fields of dimensions n and p ($p < n$), and $g(x)$, $b(x)$ are matrices of dimensions $n \times m$ and $n \times p$, respectively. In these rate-based models of the form in Eq.6.11, the singular term $(1/\epsilon)\,b(x)k(x)$ in the differential equations corresponds to the fast phenomena for which the rate expressions involve large parameters of the form $1/\epsilon$, e.g., large reaction rate and mass transfer coefficients. For a detailed description of this singular term in typical fast-rate chemical processes, see the application examples in Sections 7.1,7.2.

Clearly, the system in Eq.6.11 is not in the standard singularly perturbed form. In the following section, we analyze the two-time-scale property of the system in Eq.6.11 and address the construction of nonlinear coordinate changes that yield a standard singularly perturbed representation of the system. The material presented follows closely the results in [83]. With a slight abuse of notation, we will use the notation $L_b k(x)$ to denote a matrix whose (i, j)th component is $L_{b_j} k_i(x)$, where b_j, k_i denote the jth column and ith component of the matrix $b(x)$ and the vector field $k(x)$, respectively. Similarly, $L_f k(x)$ will be used to denote a vector field whose ith component is $L_f k_i(x)$.

6.3 Derivation of Standard Singularly Perturbed Representation of Rate-Based Models

For the system in Eq.6.11, we perform a two-time-scale decomposition to derive separate representations of its slow and fast dynamics in the appropriate time scales, and obtain some insight on the variables that should be used as part of the desired coordinate change that yields a standard singularly perturbed representation. We assume that in the system of Eq.6.11, the matrix $b(x)$ and the Jacobian $(\partial k(x)/\partial x)$ have full column and row rank p, respectively. The assumption on the rank of $b(x)$ is not restrictive. For if rank $b(x) = \bar{p} < p$, then there always exists a nonsingular $p \times p$ matrix $E(x)$ such that:

$$b(x)k(x) = b(x)E(x)E(x)^{-1}k(x) = [\bar{b}(x) \ 0] \, \bar{k}(x)$$

where $\bar{b}(x)$ has full column rank $\bar{p}$. Thus, discarding the $(p - \bar{p})$ columns that are zero and the corresponding last $(p - \bar{p})$ components of $\bar{k}(x)$, a system in the form of Eq.6.11 is obtained where the new matrix $\bar{b}(x)$ has full column rank. The condition on the rank of $(\partial k(x)/\partial x)$ is required to ensure that in the limit $\epsilon = 0$, the DAE system (see Eq.6.13) that describes the slow dynamics of Eq.6.11 has a finite index and a well-defined solution, and it is satisfied in most practical applications (see for example the chemical reactor applications studied in Section 7.1,7.2).

Consider the system of Eq.6.11 in the slow time scale t, satisfying the above assumptions. Multiplying Eq.6.11 by ϵ and considering the limit $\epsilon \to 0$, the following (linearly independent) constraints are obtained:

$$k_i(x) = 0, \quad i = 1, \ldots, p \qquad (6.12)$$

where $k_i(x)$ denotes the ith component of $k(x)$. These constraints must be satisfied in the slow system. Moreover, for the system of Eq.6.11 in the limit $\epsilon \to 0$, the term $(1/\epsilon)k(x)$ becomes indeterminate. In the fast-rate processes, this corresponds to the fact that in the limit when the large parameters in the rate expressions approach infinity, the fast mass transfer, reactions, etc., approach the quasi-steady-state conditions of phase, reaction equilibrium (specified by $k(x) = 0$), and the rates of mass transfer and reactions, as given by the explicit rate expressions, become indeterminate. It should be mentioned that, in general, these mass transfer/reaction rates, etc., are not zero.

Thus, in the limit $\epsilon \to 0$, defining $z_i = \lim_{\epsilon \to 0} (k_i(x)/\epsilon)$ as the finite but unknown rates of the fast reactions, mass transfer, etc., the system of Eq.6.11 takes the following form:

$$\dot{x} = f(x) + g(x)u + b(x)z$$
$$0 = k(x) \qquad (6.13)$$

which describes the slow dynamics of Eq.6.11. Note that the above system is a DAE system of the form in Eq.1.9 (with $l(x)$, $c(x) \equiv 0$), where x is the vector of differential variables and $z \in \mathbb{R}^p$ is the vector of algebraic variables. The above DAE system corresponds to the equilibrium-based models given in Chapter 2.

Clearly, the DAE system in Eq.6.13 has a high index ($\nu_d > 1$), and, thus, it is characterized by the presence of algebraic constraints in the differential variables x. In particular, the algebraic equations in Eq.6.13 already denote p such constraints, and depending on the index ν_d, additional constraints may also be present. These constraints specify the dimension of the subspace in $\mathcal{X}$ where the slow dynamics of the two-time-scale system of Eq.6.11 evolves.

96

On the other hand, introducing the fast time scale $\tau = t/\epsilon$ and considering the limit $\epsilon \to 0$ in the system of Eq.6.11, the following representation of the fast dynamics of Eq.6.11 is obtained:

$$\frac{dx}{d\tau} = b(x)k(x) \tag{6.14}$$

From the above system, it is clear that in the fast time scale τ, the algebraic constraints of Eq.6.12 are not satisfied, i.e., $k(x) \neq 0$. The fact that $k_i(x)$, $i = 1,\ldots,p$ are identically zero in the slow system of Eq.6.13 in the slow time scale t, while they are nonzero for the system of Eq.6.14 in the fast time scale τ, indicates that they should be used in the definition of the fast variables in the desired coordinate change.

Theorem 6.1 provides necessary and sufficient conditions for the existence of an ϵ-independent coordinate change that transforms the two-time-scale system of Eq.6.11 into a standard singularly perturbed form. For systems of Eq.6.11 where these conditions are satisfied, an explicit coordinate change is developed along the previous arguments.

Theorem 6.1: *Consider the system of Eq.6.11, for which the slow dynamics is described by the DAE system of Eq.6.13. The system of Eq.6.11 can be transformed into a two-time-scale singularly perturbed system in standard form through an ϵ-independent nonlinear coordinate change, if and only if:*

(i) the $p \times p$ matrix $L_b k(x)$ is nonsingular on $\mathcal{X}$, and

(ii) the p-dimensional distribution $B(x) = \text{span} \{b_1(x),\ldots,b_p(x)\}$ is involutive.

If these conditions hold, then under the coordinate change:

$$\begin{bmatrix} \zeta \\ \eta \end{bmatrix} = T(x) = \begin{bmatrix} \phi(x) \\ k(x) \end{bmatrix} \tag{6.15}$$

where $\zeta \in \mathbb{R}^{n-p}$, $\eta \in \mathbb{R}^p$ and $\phi(x)$ is a vector field of dimension $(n-p)$ with components $\phi_i(x)$ such that $L_{b_j}\phi_i(x) \equiv 0$, $\forall i,j$, the system of Eq.6.11 takes the following standard singularly perturbed form:

$$\begin{aligned}
\dot{\zeta} &= \tilde{f}(\zeta,\eta) + \tilde{g}(\zeta,\eta)u \\
\epsilon\dot{\eta} &= \epsilon\overline{f}(\zeta,\eta) + \epsilon\overline{g}(\zeta,\eta)u + Q(\zeta,\eta)\eta \\
y_i &= h_i(\zeta,\eta), \quad i = 1,\ldots,m
\end{aligned} \tag{6.16}$$

where $\tilde{f} = L_f\phi(x)$, $\overline{f} = L_f k(x)$, $\tilde{g} = L_g\phi(x)$, $\overline{g} = L_g k(x)$, $Q = L_b k(x)$, evaluated at $x = T^{-1}(\zeta,\eta)$, and $Q(\zeta,\eta)$ is nonsingular uniformly in ζ, η.

Proof: First, we will prove the necessity of the conditions (i) and (ii) for the existence of an ϵ-independent coordinate change that yields a representation of the system of Eq.6.11, in the standard singularly perturbed form. Thereafter, we prove the sufficiency through an explicit construction of such a coordinate change.

97

Necessity: Consider the two-time-scale system in Eq.6.11 for which there exists a nonlinear ϵ-independent coordinate change of the form:

$$\begin{bmatrix} \zeta \\ \eta \end{bmatrix} = T(x) = \begin{bmatrix} \alpha(x) \\ \beta(x) \end{bmatrix} \tag{6.17}$$

where $\zeta \in \mathbb{R}^{n-\bar{p}}$ is the vector of slow variables, $\eta \in \mathbb{R}^{\bar{p}}$ is the vector of fast variables, and $\alpha(x), \beta(x)$ are smooth vector fields of dimensions $(n-\bar{p})$ and $\bar{p}$, respectively, such that in these transformed coordinates, the system has the following representation in the standard form:

$$\begin{aligned} \dot{\zeta} &= F^1(\zeta, \eta, u, \epsilon) \\ \epsilon\dot{\eta} &= F^2(\zeta, \eta, u, \epsilon) \end{aligned} \tag{6.18}$$

where:

$$\begin{aligned} F^1(\zeta, \eta, u, \epsilon) &= [L_f\alpha(x) + \frac{1}{\epsilon}L_b\alpha(x)k(x) + L_g\alpha(x)u]_{x=T^{-1}(\zeta,\eta)} \\ F^2(\zeta, \eta, u, \epsilon) &= [\epsilon L_f\beta(x) + L_b\beta(x)k(x) + \epsilon L_g\beta(x)u]_{x=T^{-1}(\zeta,\eta)} \end{aligned} \tag{6.19}$$

Note that $\bar{p}$ is an arbitrary integer less than n. It will be shown that, in fact, $\bar{p} = p$, i.e., the two-time-scale system in Eq.6.18 has exactly p fast states and $n-p$ slow states.

For the system in Eq.6.18, the $\bar{p} \times \bar{p}$ Jacobian $(\partial F^2(\zeta, \eta, u, 0)/\partial \eta)$ is nonsingular, where:

$$F^2(\zeta, \eta, u, 0) = L_b\beta(x)k(x)|_{x=T^{-1}(\zeta,\eta)} = \begin{bmatrix} L_{b_1}\beta_1(x) & \cdots & L_{b_p}\beta_1(x) \\ \vdots & & \vdots \\ L_{b_1}\beta_{\bar{p}}(x) & \cdots & L_{b_p}\beta_{n-\bar{p}}(x) \end{bmatrix} k(x)|_{x=T^{-1}(\zeta,\eta)}$$

From our previous analysis of the slow dynamics of (Eq.6.11) in the slow time scale t, we know that $\lim_{\epsilon \to 0} k(x) = 0$, i.e., $\hat{k}(\zeta, \eta) = k(x)_{x=T^{-1}(\zeta,\eta)} = 0$. Thus, the nonsingularity of the $\bar{p} \times \bar{p}$ Jacobian:

$$\frac{\partial F^2(\zeta, \eta, u, 0)}{\partial \eta} = L_b\beta(x)|_{x=T^{-1}(\zeta,\eta)} \times \left(\frac{\partial \hat{k}(\zeta, \eta)}{\partial \eta}\right)$$

implies that the $\bar{p} \times p$, $p \times \bar{p}$ matrices $L_b\beta(x)$ and $(\partial\hat{k}(\zeta, \eta)/\partial\eta)$ have full ranks $\bar{p}$ and thus, $\bar{p} \leq p$. Furthermore, the two-time-scale system of Eq.6.18 has the following representation in the fast time scale τ:

$$\begin{aligned} \zeta' &= \epsilon F^1(\zeta, \eta, u, \epsilon) \\ \eta' &= F^2(\zeta, \eta, u, \epsilon) \end{aligned} \tag{6.20}$$

where in the limit $\epsilon \to 0$, the slow variables ζ are constant. This, together with the fact that $k(x)$ in the fast time scale τ, is an arbitrary nonzero vector field of dimension

p, leads to the conclusion that:

$$L_b\alpha(x) = \begin{bmatrix} L_{b_1}\alpha_1(x) & \cdots & L_{b_p}\alpha_1(x) \\ \vdots & & \vdots \\ L_{b_1}\alpha_{n-\bar{p}}(x) & \cdots & L_{b_p}\alpha_{n-\bar{p}}(x) \end{bmatrix} \equiv 0$$

Equivalently, the $(n - \bar{p})$ gradient covector fields $d\alpha_i(x)$, $i = 1, \ldots, (n - \bar{p})$ are orthogonal to the p-dimensional distribution $B(x) = span\{b_1(x), \ldots, b_p(x)\}$ and thus, $\bar{p} \geq p$. Clearly, in view of the previous inequality $\bar{p} \leq p$, it directly follows that $\bar{p} = p$ and the distribution $B(x)$ is involutive, from Frobenius Theorem [64]. This proves the necessity of condition (ii). Furthermore, the $p \times p$ matrices:

$$L_b\beta(x)|_{x=T^{-1}(\zeta,\eta)}\ , \quad (\frac{\partial \hat{k}(\zeta,\eta)}{\partial \eta})$$

are nonsingular. With the above observations, a proof for condition (i) can be obtained by evaluating the $p \times p$ matrix $L_b k(x) = (\partial k(x)/\partial x)b(x)$ in the transformed coordinates (ζ, η):

$$\begin{aligned} L_b k(x)|_{x=T^{-1}(\zeta,\eta)} &= (\frac{\partial \hat{k}(\zeta,\eta)}{\partial \eta})(\frac{\partial \beta(x)}{\partial x})b(x) + (\frac{\partial \hat{k}(\zeta,\eta)}{\partial \zeta})(\frac{\partial \alpha(x)}{\partial x})b(x)\Big|_{x=T^{-1}(\zeta,\eta)} \\ &= (\frac{\partial \hat{k}(\zeta,\eta)}{\partial \eta})L_b\beta(x) + (\frac{\partial \hat{k}(\zeta,\eta)}{\partial \zeta})L_b\alpha(x)\Big|_{x=T^{-1}(\zeta,\eta)} \end{aligned}$$

Clearly, from the fact that $L_b\alpha(x)|_{x=T^{-1}(\zeta,\eta)} \equiv 0$ and that the matrices $(\partial \hat{k}(\zeta,\eta)/\partial \eta)$, $L_b\beta(x)|_{x=T^{-1}(\zeta,\eta)}$ are nonsingular, it is evident that $L_b k(x)$ is also nonsingular, proving the necessity of condition (i).

Sufficiency: The sufficiency of conditions (i) and (ii) is established through the construction of a nonlinear ϵ-independent coordinate change that yields the standard form representation in Eq.6.16. More specifically, from the Frobenius Theorem, the involutivity of the distribution $B(x)$ in condition (ii) implies the existence of exactly $(n-p)$ scalar functions $\phi_i(x)$, $i = 1, \ldots, (n-p)$ such that the gradient covector fields $d\phi_i(x)$ are linearly independent and $L_{b_j}\phi_i(x) \equiv 0$, $\forall j$. Moreover, the nonsingularity of $L_b k(x)$ in the condition (i) implies that:

$$\begin{bmatrix} \zeta \\ \eta \end{bmatrix} = T(x) = \begin{bmatrix} \phi(x) \\ k(x) \end{bmatrix}$$

is a valid local diffeomorphism, where $\phi(x) = [\phi_1(x) \cdots \phi_{n-p}(x)]^T$. In these coordinates, the system of Eq.6.11 has the representation in Eq.6.16, which is in the standard form since the matrix $Q(\zeta, \eta, 0) = L_b k(x)_{x=T^{-1}(\zeta,\eta)}$ is nonsingular uniformly in ζ, η and $\eta = 0$ is the quasi-steady-state solution for the fast variables η. This completes the proof of the sufficiency. $\square$

Remark 6.2: Condition (i) of Theorem 6.1 essentially means that the corresponding DAE system in Eq.6.13 has an index two, which directly fixes the dimensions of the fast and slow variables to p and $n - p$, respectively. Condition (ii) of the theorem ensures that the $(n-p)$-dimensional slow ζ-subsystem can be made independent of the singular term $(1/\epsilon)$, thereby yielding the system of Eq.6.16 in the standard singularly perturbed form. While condition (ii) is trivially satisfied for all linear systems and for nonlinear systems with $p = 1$, it is not satisfied in general for nonlinear systems with $p > 1$.

Remark 6.3: Consider the DAE system in Eq.6.13 that describes the slow dynamics of Eq.6.11 for which condition (i) of Theorem 6.1 is satisfied, i.e., the DAE system has an index two. The DAE system has exactly p constraints and from the result in Proposition 3.3, a minimal-order state-space realization of dimension $n - p$ can be derived, using the constraints $k(x) = 0$ as part of a coordinate change. The condition (ii) of the theorem allows the derivation of such a state-space realization without evaluating a solution for the algebraic variables z, through an appropriate coordinate change (see Remark 3.2). In fact, the choice of the state variables ζ, η in Eq.6.15 precisely comprises such a coordinate change, and the description of the slow subsystem of Eq.6.16:

$$\dot{\zeta} = \widetilde{f}(\zeta, 0) + \widetilde{g}(\zeta, 0)u$$
$$y_i = h_i(x)|_{x=T^{-1}(\zeta,0)}, \quad i = 1, \ldots, m \tag{6.21}$$

is the corresponding state-space realization of the index-two DAE system in Eq.6.13.

Remark 6.4: The conditions (i) and (ii) of Theorem 6.1 are readily verifiable, and they comprise an explicit coordinate-dependent form of the abstract geometric conditions in [100] for two-time-scale systems with the specific structure of Eq.6.11. The latter point can be verified for the system of Eq.6.11 in the fast time scale:

$$\frac{dx}{d\tau} = \epsilon f(x) + \epsilon g(x)u + b(x)k(x) \tag{6.22}$$

where in the limiting case $\epsilon \to 0$, the constraints $k(x)=0$ specify the $(n-p)$-dimensional equilibrium manifold $E^u = \{x \in \mathcal{X} \ : \ k(x)=0\}$, and $C = \{x \in \mathcal{X} \ : \ \phi(x)=c \in \mathbb{R}^{(n-p)}\}$ denotes the family of p-dimensional conservation manifolds that are transversal to the equilibrium manifold E^u.

Motivated by the fact that the involutivity condition (ii) may be violated in general nonlinear systems with $p > 1$, we now consider systems of Eq.6.11 for which the distribution $B(x)$ is not involutive. For such systems, we provide a result in the following theorem which states the implications of the lack of involutivity of $B(x)$ on the two-time-scale property and the structure of the coordinate change required for obtaining a standard singularly perturbed representation.

Theorem 6.2: *Consider the system of Eq.6.11 for which the $p \times p$ matrix $L_b k(x)$ is nonsingular on $\mathcal{X}$ and the distribution $B(x)$ is not involutive. Then,*

(i) *this system exhibits two-time-scale behavior in a region $\mathcal{M}(\epsilon) \subset \mathcal{X}$, where $k_i(x) = O(\epsilon)$ for some $i \in [1, p]$, and*

(ii) *an ϵ-dependent coordinate change, singular at $\epsilon = 0$, is necessary to obtain a standard form representation of this system.*

Proof: Under condition (i) of Theorem 6.1, the system of Eq.6.11 has $n - p$ slow and p fast modes. Statement (i) is proved by contradiction. Assume that the system of Eq.6.11 exhibits a two-time-scale property in a region $\bar{\mathcal{M}} \subset \mathcal{X}$ where $k_i(x) = O(1)$, $\forall i$, and $\dim \bar{\mathcal{M}} = n$, independently of ϵ. This implies that there exists an ϵ-dependent coordinate change of the form:

$$\begin{bmatrix} \zeta \\ \eta \end{bmatrix} = T(x, \epsilon) = \begin{bmatrix} \phi(x, \epsilon) \\ \psi(x, \epsilon) \end{bmatrix} \tag{6.23}$$

which is *not* singular at $\epsilon = 0$, i.e., $T(x, 0)$ is a valid coordinate change and yields the following representation of the system of Eq.6.11:

$$\begin{aligned} \dot{\zeta} &= L_f \phi(x, \epsilon) + L_g \phi(x, \epsilon) u + \frac{1}{\epsilon} L_b \phi(x, \epsilon) k(x) \\ \epsilon \dot{\eta} &= \epsilon L_f \psi(x, \epsilon) + \epsilon L_g \psi(x, \epsilon) u + L_b \psi(x, \epsilon) k(x) \end{aligned} \tag{6.24}$$

where $x = T^{-1}(\zeta, \eta, \epsilon)$, and ζ and η are the slow and fast state vectors of dimensions $n - p$ and p, respectively. The fact that ζ_j, $j = 1, \ldots, n - p$ are the slow variables implies that in the fast time scale $\tau = t/\epsilon$, $\frac{d\zeta_j}{d\tau} = O(\epsilon)$, and thus, $L_b \phi_j(x, \epsilon) k(x) = O(\epsilon)$. However, since $B(x)$ is not involutive, at least one of the elements of the $(n - p) \times p$ matrix $L_b \phi(x, \epsilon)$ is of $O(1)$, which directly implies that $k_i(x) = O(\epsilon)$ for at least one $i \in [1, p]$, thus yielding a contradiction.

Regarding statement (ii) of the theorem, we have that the system in Eq.6.24 is clearly not in the standard form since $\dot{\zeta}$ depends explicitly on the term $1/\epsilon$, thereby establishing the necessity of a coordinate change that is singular at $\epsilon = 0$. $\qquad \square$

Remark 6.5: The results of Theorems 6.1 and 6.2 demonstrate the connection between the involutivity of $B(x)$ and the size of the state-space region in which the system of Eq.6.11 exhibits the two-time-scale property as well as the nature of the coordinate transformation required. Specifically, if $B(x)$ is involutive, the size of this state-space region is independent of ϵ, whereas if $B(x)$ is not involutive, the size of this state-space region depends explicitly on ϵ in a singular fashion and it shrinks as $\epsilon \to 0$. This is illustrated through the following simple example.

Example 6.1: Consider the system:

$$\begin{bmatrix} \dot{x}_1 \\ \dot{x}_2 \\ \dot{x}_3 \end{bmatrix} = \begin{bmatrix} 2x_1 - x_3 \\ 2x_2 + x_3 \\ x_1 + x_2 - x_3 \end{bmatrix} + \frac{1}{\epsilon} \begin{bmatrix} 1 & 1 \\ -2 & 1 \\ 2(x_1 + x_2) + 3x_3 & -1 \end{bmatrix} \begin{bmatrix} 2x_1 + x_2 \\ x_2 - 2x_1 \end{bmatrix} \qquad (6.25)$$

in the form of Eq.6.11 with $n=3$, $p=2$, $g(x)=0$ and

$$f(x) = \begin{bmatrix} 2x_1 - x_3 \\ 2x_2 + x_3 \\ x_1 + x_2 - x_3 \end{bmatrix}, \quad b(x) = \begin{bmatrix} 1 & 1 \\ -2 & 1 \\ 2(x_1 + x_2) + 3x_3 & -1 \end{bmatrix}, \quad k(x) = \begin{bmatrix} 2x_1 + x_2 \\ x_2 - 2x_1 \end{bmatrix}$$

For the above system, condition (i) of Theorem 6.1 is satisfied, i.e.,

$$L_b k(x) = \begin{bmatrix} 0 & 3 \\ -4 & -1 \end{bmatrix}$$

is nonsingular, and in the limit $\epsilon \to 0$ in the slow time scale t, the constraints of Eq.6.12 become $2x_1 + x_2 = 0$, $x_2 - 2x_1 = 0$, i.e., $x_1 = 0$ and $x_2 = 0$. Thus, the above system has two fast variables and one slow variable $\zeta = \phi(x, \epsilon)$ which must be a function of x_3, i.e., $\partial \phi(x, 0) / \partial x_3 \neq 0$. In the fast time scale $\tau = t/\epsilon$, the dynamics of such a variable ζ is described by:

$$\frac{d\zeta}{d\tau} = \epsilon L_f \phi(x, \epsilon) + L_{b_1} \phi(x, \epsilon) k_1(x) + L_{b_2} \phi(x, \epsilon) k_2(x) \qquad (6.26)$$

Clearly ζ is a slow variable if $d\zeta/d\tau = O(\epsilon)$ for an arbitrary initial condition $\zeta(0) = \phi(x(0), \epsilon)$. For this to be true, $L_{b_1} \phi(x, \epsilon) k_1(x)$ and $L_{b_2} \phi(x, \epsilon) k_2(x)$ must be $O(\epsilon)$. However, for the system in Eq.6.25, the distribution $B(x) = \text{span}\{b_1(x), b_2(x)\}$ is not involutive. Thus, either $L_{b_1} \phi(x, \epsilon)$ or $L_{b_2} \phi(x, \epsilon)$ is $O(1)$ and correspondingly $k_1(x)$ or $k_2(x)$ must be $O(\epsilon)$. In fact, for this system, if either $k_1(x(t_0))$ or $k_2(x(t_0))$ is $O(1)$ at time t_0, then both $k_1(x(t))$ and $k_2(x(t))$ become $O(1)$ at some time $t > t_0$. Thus, ζ is a slow variable only in the region where both $k_1(x)$ and $k_2(x)$ are $O(\epsilon)$, i.e., $x_1 = O(\epsilon)$ and $x_2 = O(\epsilon)$. In this region, $\zeta = x_3$ is such a slow variable. Figure 6.1 shows the solution of x_3 in the fast boundary layer for $\epsilon = 0.01$, starting from two initial conditions, (a) $x(0) = [0.02\ 0.02\ 0.3]^T$ within the above-mentioned region, and (b) $x(0) = [0.2\ 0.2\ 0.3]^T$ outside this region. Clearly, in case (a), x_3 is a slow variable and is essentially constant in the fast boundary layer, while in case (b), x_3 varies significantly in this time interval.

An ϵ-dependent coordinate change that is singular at $\epsilon = 0$ and can be used to derive a standard singularly perturbed representation of the system in Eq.6.11, can be obtained following [78] and is of the form:

$$\begin{bmatrix} \zeta \\ \eta \end{bmatrix} = T(x, \epsilon) = \begin{bmatrix} \phi(x) \\ \dfrac{k(x)}{\epsilon} \end{bmatrix} \qquad (6.27)$$

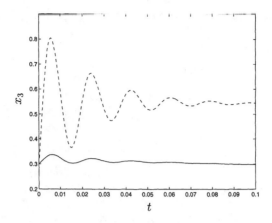

Figure 6.1: Solution of x_3 for $\epsilon = 0.01$, starting from the initial conditions $x(0) = [0.02\ 0.02\ 0.3]^T$ (solid) and $x(0) = [0.2\ 0.2\ 0.3]^T$ (dashed).

where $\zeta \in \mathbb{R}^{n-p}$ is the vector of slow variables and $\eta \in \mathbb{R}^p$ is the vector of fast variables. Under the coordinate change of Eq.6.27, the two-time-scale system of Eq.6.11 takes the following standard form:

$$\dot{\zeta} = \widetilde{f}(\zeta, \epsilon\eta) + \widetilde{g}(\zeta, \epsilon\eta)u + \widetilde{b}(\zeta, \epsilon\eta)\eta$$
$$\epsilon\dot{\eta} = \overline{f}(\zeta, \epsilon\eta) + \overline{g}(\zeta, \epsilon\eta)u + Q(\zeta, \epsilon\eta)\eta$$
$$y_i = h_i(\zeta, \epsilon\eta), \ i = 1, \dots, m \tag{6.28}$$

where $\widetilde{f} = L_f\phi(x)$, $\overline{f} = L_f k(x)$, $\widetilde{g} = L_g\phi(x)$, $\overline{g} = L_g k(x)$, $Q = L_b k(x)$, $\widetilde{b} = L_b\phi(x)$, evaluated at $x = T^{-1}(\zeta, \epsilon\eta)$, and $Q(\zeta, 0)$ is nonsingular uniformly in ζ.

The coordinate change of Eq.6.27 essentially defines the fast variables η as the constraints $k(x)$ "stretched" by a factor $1/\epsilon$, thereby making the ϵ-dependent region $\mathcal{M}(\epsilon)$ of Theorem 6.2, independent of ϵ in the transformed coordinates (ζ, η). However, such a coordinate change may be unnecessarily restrictive in the sense that the resulting standard singularly perturbed representation of Eq.6.28 is valid only in a subspace region $\bar{\mathcal{M}}(\epsilon) = \{x \in \mathcal{X} : k_i(x) = O(\epsilon), \text{ for all } i\}$. Under suitable conditions, a coordinate change that "stretches" a minimal number of fast variables by the factor $1/\epsilon$ can be used, thereby yielding a standard singularly perturbed representation of the system of Eq.6.11 which is valid in a larger region. The result is given in the following proposition.

Proposition 6.1: *Consider the system of Eq.6.11, for which the $p \times p$ matrix $L_b k(x)$ is nonsingular on $\mathcal{X}$ and the distribution $B(x)$ is not involutive. For this system, let $\bar{p} < p$ be the largest integer such that, after possibly rearranging the columns of $b(x)$ and correspondingly the rows of $k(x)$,*

103

(i) the distribution $\bar{B}(x) = span\{b_1(x), \ldots, b_{\bar{p}}(x)\}$ is involutive, and

(ii) $L_{b_j} k_i(x) \equiv 0;\ \ i = \bar{p}+1, \ldots, p,\ \ j = 1, \ldots, \bar{p}.$

Then, under the coordinate change:

$$
\begin{bmatrix} \zeta \\ \eta_1 \\ \eta_2 \end{bmatrix} = T(x, \epsilon) = \begin{bmatrix} \phi(x) \\ k^1(x) \\ k^2(x) \\ \epsilon \end{bmatrix}
\tag{6.29}
$$

where $\zeta \in \mathbb{R}^{n-p}$, $\eta_1 \in \mathbb{R}^{\bar{p}}$, $\eta_2 \in \mathbb{R}^{p-\bar{p}}$, $k^1(x), k^2(x)$ are vector fields comprised of the first $\bar{p}$ and last $p - \bar{p}$ components of $k(x)$, respectively, and $\phi(x)$ is a vector field of dimension $(n - p)$ with components $\phi_i(x)$ such that $L_{b_j}\phi_i(x) \equiv 0,\ j = 1, \ldots, \bar{p},\ \forall i$, the system of Eq.6.11 takes the following standard singularly perturbed form:

$$
\begin{aligned}
\dot{\zeta} &= \tilde{f}(\zeta, \eta_1, \epsilon\eta_2) + \tilde{g}(\zeta, \eta_1, \epsilon\eta_2)u + \tilde{b}(\zeta, \eta_1, \epsilon\eta_2)\eta_2 \\
\epsilon\dot{\eta}_1 &= \epsilon\tilde{f}_1(\zeta, \eta_1, \epsilon\eta_2) + \epsilon\bar{g}_1(\zeta, \eta_1, \epsilon\eta_2)u + Q_{11}(\zeta, \eta_1, \epsilon\eta_2)\eta_1 + \epsilon Q_{12}(\zeta, \eta_1, \epsilon\eta_2)\eta_2 \\
\epsilon\dot{\eta}_2 &= \bar{f}_2(\zeta, \eta_1, \epsilon\eta_2) + \bar{g}_2(\zeta, \eta_1, \epsilon\eta_2)u + Q_{22}(\zeta, \eta_1, \epsilon\eta_2)\eta_2 \\
y_i &= h_i(\zeta, \eta_1, \epsilon\eta_2),\ i = 1, \ldots, m
\end{aligned}
\tag{6.30}
$$

where $\tilde{f} = L_f\phi(x)$, $\tilde{g} = L_g\phi(x)$, $\tilde{b} = L_{b^2}\phi(x)$, $\bar{f}_i = L_f k^i(x)$, $\bar{g}_i = L_g k^i(x)$, $Q_{11} = L_{b^1} k^1(x)$, $Q_{12} = L_{b^2} k^1(x)$, $Q_{22} = L_{b^2} k^2(x)$ evaluated at $x = T^{-1}(\zeta, \eta_1, \epsilon\eta_2)$, and $b^1(x), b^2(x)$ denote the matrices formed by the first $\bar{p}$ and last $p - \bar{p}$ columns of $b(x)$, respectively.

Proof: A proof of the above proposition follows from a straightforward derivation of the representation in Eq.6.30 in the new coordinates given in Eq.6.29. The system of Eq.6.30 is in standard form owing to the fact that the matrix:

$$
\begin{bmatrix} Q_{11} & Q_{12} \\ 0 & Q_{22} \end{bmatrix} = L_b k(x)
$$

is nonsingular. The quasi-steady-state solution for η_1, η_2 are given by $\eta_1 = 0$ and $\eta_2 = -Q_{22}^{-1}[\bar{f}_2(\zeta, 0, 0) + \bar{g}_2(\zeta, 0, 0)]$ $\qquad\square$

Remark 6.6: For the two-time-scale system in Eq.6.11, the singularly perturbed representations in Eqs.6.16, 6.28, and 6.30 are in standard form if the matrix $L_b k(x)$ is nonsingular on $\mathcal{X}$. Furthermore, it can be verified that in all these cases, the fast subsystem is exponentially stable if $L_b k(x)$ is Hurwitz uniformly in $x \in \mathcal{X}$. This is true for most fast-rate processes (see, e.g., the applications studied in Sections 7.1, 7.2), and thus, a feedback controller for the process can be designed on the basis of the slow subsystem, ignoring the stable fast dynamics. Furthermore, the representations of the slow subsystems are exactly the minimal-order state-space realization of the DAE system in Eq.6.13, in the coordinates ζ, η. These observations provide a rigorous justification for using the "equilibrium-based" DAE models given in Chapter 2, as the basis for controller synthesis for the fast-rate processes.

6.4 Singularly Perturbed Models of Process Networks

Until now we focused on the class of singularly perturbed systems of Eq.6.1 that arise as detailed rate-based dynamic models of chemical processes with fast heat/mass transfer, reactions, etc. For such singularly perturbed systems, the results of the previous section apply if, in the limit $\epsilon \to 0$, the corresponding DAE system of Eq.6.13 has an index two. This condition is indeed satisfied for process units with fast heat/mass transfer, reactions, etc. (see the examples in Chapter 2).

However, for process networks consisting of several units with large recycle streams (see the example in Section 2.5), the DAE system that describes the slow dynamics of the overall network may have an index exceeding two. Furthermore, unlike the fast-rate processes where the singular perturbation parameter ϵ is the reciprocal of a large parameter, the parameter ϵ in process networks is the reciprocal of the large recycle flow rate. Typically, these large flow rates are manipulated inputs in a control problem. Thus, the results of previous section do not apply to the models of such process networks.

For process networks with a large recycle, defining ϵ as the reciprocal of the large recycle flow rate, the overall dynamic model has the following general form:

$$\dot{x} = f(x) + g(x)u^s + \frac{1}{\epsilon} b(x) c(x) u^l$$
$$y_i = h_i(x) \ i = 1, \ldots, m \qquad (6.31)$$

where $u^s \in \mathbb{R}^{m_1}$ and $u^l \in \mathbb{R}^{m_2}$ are vectors of manipulated inputs corresponding to the small and large flow rates in the network, respectively. In a more general process network, where the recycle flow rates are large *and* the individual process units have fast and slow reactions, mass transfer, heat transfer, etc., in appropriate time scales, the corresponding model is given by the following system:

$$\dot{x} = f(x) + g(x)u^s + \frac{1}{\epsilon} b(x) \left\{ k(x) + c(x) u^l \right\}$$
$$y_i = h_i(x) \ i = 1, \ldots, m \qquad (6.32)$$

A similar time-scale decomposition can be performed for the singularly perturbed system of Eq.6.32. More specifically, considering the limit $\epsilon \to 0$ in the slow time scale t yields the following algebraic constraints $k(x) + c(x)u^l = 0$. Thus, defining the algebraic variables $z = \lim_{\epsilon \to 0} (k(x) + c(x)u^l)/\epsilon$, the corresponding slow dynamics is

105

described by the DAE system:

$$\dot{x} = f(x) + g(x)u^s + b(x)z$$
$$0 = k(x) + c(x)u^l$$
$$y_i = h_i(x) \ i = 1, \ldots, m \tag{6.33}$$

Clearly, the above DAE system has a high index. Moreover, unlike the DAE system of Eq.6.13, the above system is nonregular, i.e., the constrained state space where x evolves, is control-dependent. In the fast time scale $\tau = t/\epsilon$, the system of Eq.6.32 has the following representation in the limit $\epsilon \to 0$:

$$\frac{dx}{d\tau} = b(x)\left\{k(x) + c(x)u^l\right\} \tag{6.34}$$

for which the equilibrium manifold is clearly control-dependent.

The results of previous sections clearly do not apply to the system of Eq.6.32. The key issues that are different and need to be addressed in the analysis and the derivation of two-time-scale (or possibly even multiple-time-scale) standard form singularly perturbed representations of such systems are: (i) the possibility of an index $\nu_d > 2$ for the corresponding DAE system of Eq.6.33, and (ii) the nonregularity of the DAE system, or equivalently, the fact that the equilibrium manifold for the system of Eq.6.34 is control-dependent.

Notation

Roman letters

b, g = matrices in system description
C = characteristic matrix
f, k = vector fields in system description
$\bar{f}, \tilde{f}$ = vector fields in standard form description
$\tilde{b}, \bar{g}, \tilde{g}, Q$ = matrices in standard form description
h_i = scalar functions
$\mathcal{M}$ = subspace where two-time-scale behavior holds
m = number of manipulated inputs and controlled outputs
n = number of differential variables
p = number of algebraic variables
r_i = relative order of output y_i with respect to u
u = manipulated input vector
v = external input vector
x = vector of differential/state variables
y_i = output variable
y_{sp} = vector of output setpoints
z = vector of algebraic variables

Greek letters

ϵ = singular perturbation parameter
β_{ij}, γ_{ij} = vectors of adjustable controller parameters
ζ = vector of slow variables in two-time-scale system
η = vector of fast variables in two-time-scale system
ν_d = index of a DAE system
ϕ_i = scalar functions for coordinate transformation

Math symbols

$\mathbb{R}$ = real line
$\mathbb{R}^i$ = i-dimensional Euclidean space
$[\,\cdot\,]^T$ = transpose of a vector/matrix
$L_f \alpha(x)$ = Lie derivative of a scalar function $\alpha(x)$ with respect to

vector field $f(x)$, defined as $L_f \alpha(x) = [\dfrac{\partial \alpha}{\partial x_1} \cdots \dfrac{\partial \alpha}{\partial x_n}] f(x)$

$L_f^j \alpha(x)$ = high-order Lie derivative defined as $L_f^j \alpha(x) = L_f(L_f^{j-1} \alpha(x))$

$[f_1, f_2]$ = Lie bracket of two vector fields $f_1(x)$ and $f_2(x)$, defined as

$$[f_1, f_2] = \frac{\partial f_2(x)}{\partial x} f_1(x) - \frac{\partial f_1(x)}{\partial x} f_2(x)$$

7. Simulation Studies of Chemical Process Applications

In this chapter, we study, through simulations, the control of several chemical process applications that are modeled by high-index DAEs. While these processes can also be modeled by index-one DAEs or ODEs by appropriately modifying the modeling assumptions, we illustrate the limitations of controllers designed on the basis of such models. The first two chemical reactor examples illustrate the problems in standard input/output linearizing controllers designed without accounting for the time-scale multiplicity, and the singular perturbation modeling framework of Chapter 6 for designing controllers that give good performance with stability [83]. In the first example, we illustrate the problem of controller ill-conditioning, where the control action evaluated by the controller designed on the basis of the rate-based ODE model is highly sensitive to small modeling errors. The second example illustrates the problem of closed-loop instability for slightly nonminimum phase systems, where the zero dynamics of the process is also a two-time-scale system with the fast subsystem being unstable. Both of these problems are overcome by designing the controller on the basis of a model for the slow dynamics, ignoring the fast dynamics that are stable. The reduced-order model for the slow dynamics is precisely a state-space realization of the high-index, equilibrium-based DAE model. In the third example, we illustrate through a two-phase reactor, the importance of modeling the nonreactive vapor holdup for a reliable description of the process dynamics; this is particularly relevant in the modeling and control of reactive distillation columns [86, 92]. The resulting model is a regular DAE system of index two, while a standard model obtained under the assumption of negligible vapor holdup is an index-one DAE system. The fourth example is a reactor-condenser network [88], where the slow dynamics is described by an index-three DAE system that is nonregular. For this process, we address the design of a feedforward/feedback controller to compensate for the effects of disturbances.

7.1 A CSTR with heating jacket

Consider the CSTR with heating jacket shown in Figure 2.1. Reactant A is fed to the reactor at a flow rate F_A, molar concentration C_{Ao}, and temperature T_A. An irreversible endothermic reaction $A \to B$ yields the product B, and the product stream is withdrawn at a flow rate $F_o = F_A$, i.e., the reactor holdup volume V is constant. The reaction rate R_A is given by the expression:

$$R_A = k_o e^{(-E/RT)} C_A$$

where k_o and E are the reaction rate coefficient and activation energy, respectively, T is the reactor temperature, and C_A is the molar concentration of A in the reactor. Heat is provided to the reactor from the jacket, where a heating fluid is fed at a flow rate F_h and a temperature T_h. The modeling equations for the process include the mole balances for the two components in the reactor and the enthalpy balances in the reactor and the jacket. The resulting dynamic model is given by:

$$\dot{C}_A = \frac{F_A}{V}(C_{Ao} - C_A) - k_o e^{(-E/RT)} C_A$$

$$\dot{C}_B = -\frac{F_A}{V}C_B + k_o e^{(-E/RT)} C_A$$

$$\dot{T} = \frac{F_A}{V}(T_A - T) - k_o e^{(-E/RT)} C_A \frac{\Delta H_r}{\rho c_p} + \frac{UA}{\rho c_p}(\frac{T_j - T}{V})$$

$$\dot{T}_j = \frac{F_h}{V_h}(T_h - T_j) - \frac{UA}{\rho_h c_{ph}}(\frac{T_j - T}{V_h}) \qquad (7.1)$$

For simplicity, it is assumed that the density and specific heat capacities of the two liquids are the same, i.e., $\rho_h = \rho$ and $c_{ph} = c_p$, and the liquid holdup in the jacket at a temperature T_j has a constant volume V_h. Furthermore, we consider the case when the heat transfer between the jacket and the reactor, $Q = UA(T_j - T)$, is fast compared to the reaction, i.e., $(UA/\rho c_p) = (1/\epsilon)$ is a large parameter. The values of the process parameters and variables at the nominal steady state are given in Table 7.1. Note that the dynamic model in Eq.7.1 includes the rate expression for the fast heat transfer Q.

In this process, it is desired to control the product concentration C_B and reactor temperature T using the reactant flow rate F_A and the heating fluid flow rate F_h as the manipulated inputs. Under the above conditions, the system of Eq.7.1 takes the form of Eq.6.11 with the state vector $x = [C_A \ C_B \ T \ T_j]^T$, manipulated input vector $u = [F_A \ F_h]^T$, the controlled outputs $y_1 = x_2$, $y_2 = x_3$,

$$f(x) = \begin{bmatrix} -k_o e^{(-E/Rx_3)} x_1 \\ k_o e^{(-E/Rx_3)} x_1 \\ -k_o e^{(-E/Rx_3)} x_1 (\frac{\Delta H_r}{\rho c_p}) \\ 0 \end{bmatrix}, \ g(x) = \begin{bmatrix} \frac{(C_{Ao} - x_1)}{V} & 0 \\ -\frac{x_2}{V} & 0 \\ \frac{(T_A - x_3)}{V} & 0 \\ 0 & \frac{(T_h - x_4)}{V_h} \end{bmatrix}, \ b(x) = \begin{bmatrix} 0 \\ 0 \\ \frac{1}{V} \\ -\frac{1}{V_h} \end{bmatrix}$$

and $k(x) = x_4 - x_3$.

109

Table 7.1: Nominal values of variables for reactor with heating jacket

Variable	Description	Nominal value
C_{Ao}	feed reactant concentration (mol/l)	5.0
C_A	reactant concentration in reactor (mol/l)	1.596
C_B	product concentration in reactor (mol/l)	3.404
c_p	specific heat capacity $(J/g\ K)$	6.0
E	activation energy $(J/mol\ K)$	50000
F_A	outlet flow rate from reactor (l/min)	3.0
F_h	heating fluid flow rate (l/min)	0.1
k_o	pre-exponential factor in reaction rate $(l/mol\ min)$	1.0×10^9
T_A	feed reactant temperature (K)	300
T	reactor temperature (K)	284.08
T_h	heating fluid temperature (K)	375
T_j	jacket temperature (K)	285.37
V	reactor holdup volume (l)	10.0
V_h	jacket volume (l)	1.0
ρ	liquid density (g/l)	600
ΔH_r	heat of reaction (J/mol)	20000
ϵ	small parameter $(\rho c_p/UA)$ (min/l)	0.144

We first address the control of this process within a standard input/output linearization framework, ignoring its two-time-scale behavior. For the rate-based model in Eq.7.1, the relative orders of the two outputs y_1 and y_2, with respect to the manipulated input vector u, are $r_1 = 1$ and $r_2 = 1$, respectively. However, both outputs are affected more directly by the same input u_1, and thus, the characteristic matrix is singular [40]. This implies the need for a dynamic feedback controller, which is designed through dynamic extension by defining $v_1 = \dot{u}_1$ as a new manipulated input to obtain a system with an extended state vector $\bar{x} = [x^T\ u_1]^T$, new input vector $v = [\dot{u}_1\ u_2]^T$, relative orders $r_1 = 2$, $r_2 = 2$, and a characteristic matrix that is nonsingular (the reader may refer to [64] for details of the dynamic extension procedure). An input/output linearizing dynamic state feedback controller can be designed on the basis of the extended system to induce a well-characterized closed-loop response. However, owing to the fact that the relative orders of both outputs in the extended system are two, the resulting controller involves terms multiplied by the large factors $(1/\epsilon)$, $(1/\epsilon^2)$, and thus, it is severely ill-conditioned; this is illustrated through simulations later in this section (see Figures 7.2, 7.3).

We now address the control of the process, accounting for its two-time-scale behavior within the singular perturbation modeling framework of Chapter 6. It is clear that in the limit $\epsilon \to 0$, the heat transfer resistance becomes negligible and the reactor and jacket approach thermal equilibrium, i.e., $T_j \to T$. In this limiting case, the heat transfer rate Q, as given by the rate expression $Q=UA(T_j-T)$, becomes indeterminate; it is governed by the thermal equilibrium relation $T_j=T$ instead. It should be emphasized that the quasi-steady-state approximation of thermal equilibrium $T_j=T$, in general, does not imply that the heat transfer rate Q is zero. Thus, the (unknown) heat transfer rate Q is defined as the algebraic variable $z = \lim_{\epsilon\to 0}(T_j-T)/\epsilon$, to obtain a DAE representation of the slow dynamics of the form in Eq.6.13, for which $L_b k(x)$ is a scalar term that is nonzero. Furthermore, the one-dimensional distribution $B(x)=span\{b(x)\}$ is trivially involutive. Thus, the coordinate change of Theorem 6.1 takes the form:

$$
\begin{bmatrix} \zeta_1 \\ \zeta_2 \\ \zeta_3 \\ \eta \end{bmatrix} = T(x) = \begin{bmatrix} x_1 \\ x_2 \\ V x_3 + V_h x_4 \\ x_4 - x_3 \end{bmatrix}
$$

In these new coordinates ζ, η, the two-time-scale system of Eq.7.1 has the following standard singularly perturbed form:

$$\dot\zeta_1 = -R_A(\zeta,\eta) + \frac{(C_{Ao}-\zeta_1)}{V}u_1$$

$$\dot\zeta_2 = R_A(\zeta,\eta) - \frac{\zeta_2}{V}u_1$$

$$\dot\zeta_3 = -R_A(\zeta,\eta)(\frac{V\Delta H_r}{\rho c_p}) + \left(T_A - \frac{\zeta_3 - V_h\eta}{V+V_h}\right)u_1 + \left(T_h - \frac{\zeta_3 + V\eta}{V+V_h}\right)u_2$$

$$\epsilon\dot\eta = \epsilon\left\{ R_A(\zeta,\eta)(\frac{\Delta H_r}{\rho c_p}) - \frac{1}{V}\left(T_A - \frac{\zeta_3 - V_h\eta}{V+V_h}\right)u_1 + \frac{1}{V_h}\left(T_h - \frac{\zeta_3 + V\eta}{V+V_h}\right)u_2 \right\}$$
$$\qquad - \left(\frac{1}{V_h} + \frac{1}{V}\right)\eta$$

$$y_1 = \zeta_2$$

$$y_2 = \frac{\zeta_3 - V_h\eta}{V+V_h} \tag{7.2}$$

where, ζ_1, ζ_2, ζ_3 denote the slow variables, η is the fast variable and

$$R_A(\zeta,\eta) = k_o e^{(-E(V+Vh)/R(\zeta_3-V_h\eta))}\, \zeta_1$$

Note that the controlled output y_2 explicitly depends on the fast variable η, i.e., the small difference between the jacket and reactor temperatures. For the two-time-scale system of Eq.7.2, the one-dimensional fast subsystem is obtained by setting $\epsilon=0$ in the fast time scale τ:

$$\frac{d\eta}{d\tau} = -\left(\frac{1}{V_h} + \frac{1}{V}\right)\eta \tag{7.3}$$

111

which is clearly exponentially stable at $\eta = 0$ since V_h, $V > 0$. Thus, the fast and stable subsystem can be ignored and a controller can be designed on the basis of the three-dimensional slow subsystem, obtained by setting $\epsilon = 0$ in the slow time scale t:

$$\dot{\zeta}_1 = -R_A(\zeta, 0) + \frac{(C_{Ao} - \zeta_1)}{V} u_1$$

$$\dot{\zeta}_2 = R_A(\zeta, 0) - \frac{\zeta_2}{V} u_1$$

$$\dot{\zeta}_3 = -R_A(\zeta, 0)(\frac{V \Delta H_r}{\rho c_p}) + \left(T_A - \frac{\zeta_3}{V + V_h} \right) u_1 + \left(T_h - \frac{\zeta_3}{V + V_h} \right) u_2$$

$$y_{1s} = \zeta_2$$

$$y_{2s} = \frac{\zeta_3}{V + V_h} \tag{7.4}$$

Note that in the slow subsystem, $\eta \equiv 0$, i.e., $T_j = T$ – a thermal equilibrium condition, analogous to phase equilibrium conditions in multi-phase systems with fast inter-phase mass transfer. Furthermore, in the slow subsystem, the controlled output $y_{1s} = y_1$, whereas $y_{2s} = (\zeta_3/(V + V_h))$ approximates the true reactor temperature T with the weighted average of the reactor and jacket temperatures, which is in accordance with the thermal equilibrium condition. It can be verified that in the slow subsystem of Eq.7.4, the relative orders of y_{1s} and y_{2s} with respect to the input vector u are $r_1 = 1$ and $r_2 = 1$, respectively, the characteristic matrix is nonsingular, and the one-dimensional zero dynamics of the slow subsystem is exponentially stable. Thus, an input/output linearizing controller was designed and coupled with an external linear controller with integral action of the form in Eq.3.47, to induce the following decoupled, first-order response in the nominal slow subsystem:

$$y_{is} + \beta_{i1}^i \dot{y}_{is} = y_{isp}, \quad i = 1, 2 \tag{7.5}$$

where $\beta_{11}^1 = 10$ min and $\beta_{21}^2 = 15$ min. The performance and robustness of this controller was compared with that of an analogous controller synthesized on the basis of the rate-based model in Eq.7.1 to induce the following decoupled, second-order responses:

$$y_i + \beta_{i1}^i \dot{y}_i + \beta_{i2}^i \ddot{y}_i = y_{isp}, \quad i = 1, 2 \tag{7.6}$$

The controller parameters were tuned for critically-damped responses with the same time constants as in the response of Eq.7.5, i.e. $\beta_{11}^1 = 20$, $\beta_{12}^1 = 100$, $\beta_{21}^2 = 30$, $\beta_{22}^2 = 225$.

The first run was used to study the performance of the two controllers in the nominal process, for a 15% increase in the setpoint for y_1. The corresponding profiles for the controlled outputs and the manipulated inputs are shown in Figure 7.1. Clearly, the controller designed on the basis of the rate-based model in Eq.7.1 yields the desired second-order decoupled responses in the nominal process. The controller based on the slow subsystem in Eq.7.4 also yields excellent performance.

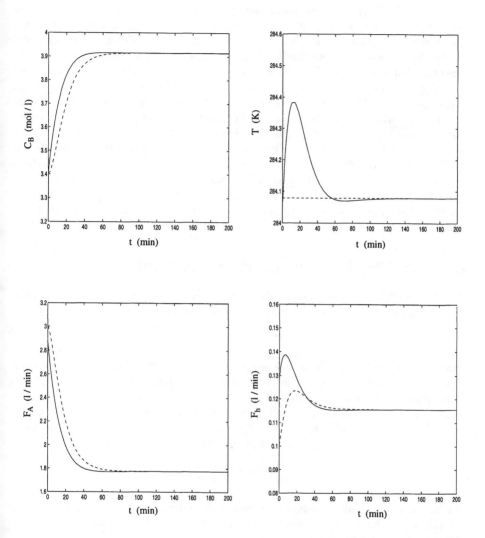

Figure 7.1: Comparison of closed-loop profiles for setpoint tracking in the nominal system, under controllers designed on the basis of the rate-based model (dashed) and the reduced-order slow subsystem (solid)

The second run was performed for the same setpoint change, in the presence of a very small (0.5%) error in the process parameters ΔH_r, ρ, and c_p. The corresponding input and output profiles under the two controllers are shown in Figure 7.2. Clearly, the controller designed on the basis of the slow subsystem yields a performance virtually indistinguishable from that in the nominal case (compare with Figure 7.1). On the other hand, the controller designed on the basis of the rate based model is very sensitive to these errors, since the effects of the small errors are magnified through the large factors $(1/\epsilon)$, $(1/\epsilon^2)$ in the control law; note the significant difference in the calculated control action (especially for F_h) compared to the nominal case, even for such a small modeling error. Figure 7.3 shows a comparison of the performances of the two controllers for a more realistic error of 5% in the same parameters. Clearly, while the controller designed on the basis of the slow subsystem still yields excellent performance, the controller designed on the basis of the rate-based ODE model leads to instability. These simulations clearly illustrate the problem of ill-conditioning in standard inversion-based controllers designed directly on the basis of the rate-based model, ignoring the time scale-multiplicity of the process.

7.2 A CSTR with multiple reactions

Consider a CSTR where reactant A is fed at a flow rate F_A, molar concentration C_{Ai} and temperature T_i, and the following elementary reactions occur in series:

$$A \rightleftharpoons B \rightarrow C \rightarrow D$$

with B as the desired product. The reaction rate for the reversible reaction $A \rightleftharpoons B$ is given by $R_1 = k_1(C_A - (C_B/\kappa_1))$ where k_1 and κ_1 are the reaction rate and equilibrium constants at the reactor temperature T, and C_A and C_B are the molar concentrations of A and B, respectively. The reaction rates for the irreversible reactions $B \rightarrow C$ and $C \rightarrow D$ are given by $R_2 = k_2 C_B$, and $R_3 = k_3 C_C$, respectively, and the three reaction rate coefficients are given by the Arrhenius relation:

$$k_i = k_i^0 \, exp \, (\frac{E_i}{R}(\frac{1}{T^0} - \frac{1}{T})) \, , \quad i = 1, 2, 3$$

where the superscript "0" refers to values at the reference temperature T^0 taken to be the reactor temperature at the nominal steady state. The dynamic model of the process has the following form:

$$\dot{C}_A = \frac{F_A}{V}(C_{Ai} - C_A) - k_1(C_A - \frac{C_B}{\kappa_1})$$

$$\dot{C}_B = -\frac{F_A}{V}C_B + k_1(C_A - \frac{C_B}{\kappa_1}) - k_2 C_B$$

$$\dot{C}_C = -\frac{F_A}{V}C_C + k_2 C_B - k_3 C_C \qquad (7.7)$$

114

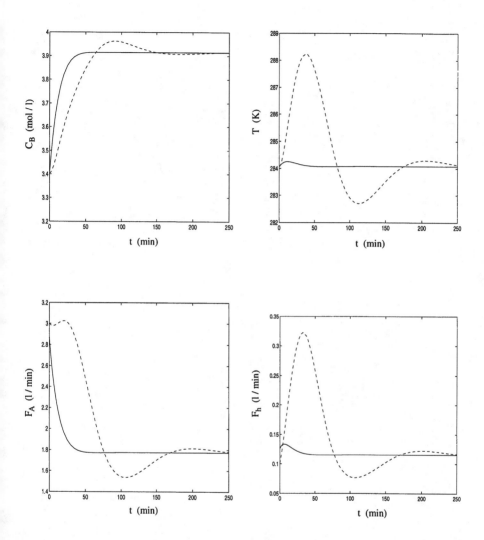

Figure 7.2: Comparison of closed-loop profiles for setpoint tracking in the presence of 0.5% error in ΔH_r, ρ, and c_p, under controllers designed on the basis of the rate-based model (dashed) and the reduced-order slow subsystem (solid)

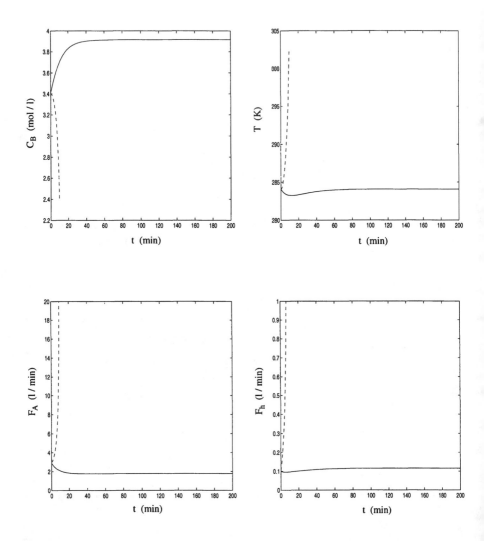

Figure 7.3: Comparison of closed-loop profiles for setpoint tracking in the presence of 5.0% error in ΔH_r, ρ, and c_p, under controllers designed on the basis of the rate-based model (dashed) and the reduced-order slow subsystem (solid)

$$\dot{C}_D = -\frac{F_A}{V}C_D + k_3 C_C$$

$$\dot{T} = \frac{F_A}{V}(T_i - T) - k_1\left(C_A - \frac{C_B}{\kappa_1}\right)\frac{\Delta H_{r1}}{\rho c_p} - k_2 C_B \frac{\Delta H_{r2}}{\rho c_p} - k_3 C_C \frac{\Delta H_{r3}}{\rho c_p} - \frac{Q}{\rho V c_p}$$

where V, ρ, and c_p are the volume, density, and specific heat capacity of the liquid holdup, assumed constant. The heats of reaction at a temperature T are given by:

$$\Delta H_{ri} = \Delta H_{ri}^0 + \Delta c_{pi}(T - T^0), \quad i = 1, 2, 3$$

where $\Delta c_{p1} = (c_{pB} - c_{pA})$, $\Delta c_{p2} = (c_{pC} - c_{pB})$, $\Delta c_{p3} = (c_{pD} - c_{pC})$, and $c_{pA}, c_{pB}, c_{pC}, c_{pD}$ are the molar heat capacities of the respective components. Furthermore, the reaction equilibrium constant κ_1 at the temperature T is governed by the following relation:

$$ln\left(\frac{\kappa_1}{\kappa_1^0}\right) = \left(\frac{\Delta H_{r1}^0 - \Delta c_{p1} T^0}{R}\right)\left(\frac{1}{T^0} - \frac{1}{T}\right) + \left(\frac{\Delta c_{p1}}{R}\right) ln\left(\frac{T}{T^0}\right)$$

For this process, it is desired to control the product concentration C_B and the reactor temperature T, using the reactant feed flow rate F_A and the heat duty Q as the manipulated inputs. Moreover, we consider the case when the first and third reactions $A \rightleftharpoons B$, $C \rightarrow D$, are much faster than the second reaction $B \rightarrow C$, i.e., $k_1^0 \gg k_2^0$ and $k_3^0 \gg k_2^0$. The nominal values of the process parameters and variables are given in Table 7.2 and correspond to a stable steady state. Defining the small parameter $\epsilon = (1/k_1^0)$ and an additional parameter $\kappa_2 = (k_3^0/k_1^0)$ where $\kappa_2 = O(1)$, the rate-based model in Eq.7.7 takes the form of the system in Eq.6.11 with the state vector $x = [C_A \; C_B \; C_C \; C_D \; T]^T$, manipulated input vector $u = [F_A \; Q]^T$, the controlled outputs $y_1 = x_2$, $y_2 = x_5$, and

$$f(x) = \begin{bmatrix} 0 \\ -k_2 x_2 \\ k_2 x_2 \\ 0 \\ -\frac{k_2 x_2 \Delta H_{r2}}{\rho c_p} \end{bmatrix}, \quad k(x) = \begin{bmatrix} e^{\frac{E_1}{R}\left(\frac{1}{T^0} - \frac{1}{x_5}\right)}\left(x_1 - \frac{x_2}{\kappa_1}\right) \\ e^{\frac{E_3}{R}\left(\frac{1}{T^0} - \frac{1}{x_5}\right)} x_3 \end{bmatrix}$$

$$g(x) = \begin{bmatrix} \frac{C_{Ai} - x_1}{V} & 0 \\ -\frac{x_2}{V} & 0 \\ -\frac{x_3}{V} & 0 \\ -\frac{x_4}{V} & 0 \\ \frac{T_i - T}{V} & -\frac{1}{\rho V c_p} \end{bmatrix}, \quad b(x) = \begin{bmatrix} -1 & 0 \\ 1 & 0 \\ 0 & -\kappa_2 \\ 0 & \kappa_2 \\ -\frac{\Delta H_{r1}}{\rho c_p} & -\frac{\kappa_2 \Delta H_{r3}}{\rho c_p} \end{bmatrix}$$

Ignoring the two-time-scale behavior of the process, an input/output linearizing controller can be designed directly on the basis of the rate-based model in Eq.7.7. The relative orders of the two outputs y_1 and y_2 with respect to the manipulated input vector u are $r_1 = 1$ and $r_2 = 1$, and the characteristic matrix is nonsingular.

117

Table 7.2: Nominal values of variables for reactor with multiple reactions

C_{Ai}	10.0 mol/l	C_A	2.333 mol/l
C_B	5.111 mol/l	C_C	0.04 mol/l
C_D	2.515 mol/l	c_p	5.0 kJ/kg K
c_{pA}	120 J/mol K	c_{pB}	80 J/mol K
c_{pC}	70 J/mol K	c_{pD}	140 J/mol K
E_1	45 kJ/mol	E_2	35 kJ/mol
E_3	40 kJ/mol	F_A	4.0 l/min
k_1^0	50 min^{-1}	k_2^0	0.2 min^{-1}
k_3^0	25 min^{-1}	T	300 K
T_i	300 K	Q	295.97 kJ/min
V	10 l	κ_1^0	2.25
ρ	0.8 kg/l	ΔH_{r1}^0	-7.0 kJ/mol
ΔH_{r2}^0	-5.0 kJ/mol	ΔH_{r3}^0	-3.0 kJ/mol

However, the three-dimensional zero dynamics is a two-time-scale system in itself, with two slow modes that are stable and a fast mode that is unstable, i.e., the process is slightly nonminimum phase [66]. This is illustrated in Figure 7.4, which shows an inverse response in y_1 in the fast boundary layer for a step increase in u_1, starting from the nominal steady state. Owing to this slightly nonminimum phase behavior, the above-mentioned controller leads to closed-loop instability, implying the need for addressing the two-time-scale behavior in the design of a well-conditioned controller that yields a good performance with stability.

For the two-time-scale system, it is clear that as $\epsilon \to 0$, $k(x) \to [0\ 0]^T$, i.e., $C_A \to (C_B/\kappa_1)$ and $C_C \to 0$. This is consistent with the physical intuition for processes with fast reactions, where reversible ones are essentially at equilibrium, and the irreversible ones proceed to complete conversion of the reactants. In this limiting case, the rates for the two fast reactions as given by the respective rate expressions, become indeterminate and are instead governed by the corresponding quasi-steady-state conditions. With this observation, the reaction rates for the two fast reactions are defined as the algebraic variables:

$$z_1 = \lim_{\epsilon \to 0} \frac{1}{\epsilon} e^{\frac{E_1}{R}(\frac{1}{T^0} - \frac{1}{x_5})} \left(x_1 - \frac{x_2}{\kappa_1}\right)$$
$$z_2 = \lim_{\epsilon \to 0} \frac{1}{\epsilon} e^{\frac{E_3}{R}(\frac{1}{T^0} - \frac{1}{x_5})} x_3$$

to obtain a DAE representation of the slow dynamics, in the form of Eq.6.13, where the matrix $L_b k(x)$ is nonsingular. However, the two-dimensional distribution $B(x) = span\{b_1(x), b_2(x)\}$ is not involutive due to the fact that the heats of reaction ΔH_{r1}

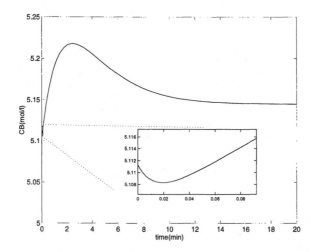

Figure 7.4: Inverse response in C_B for step increase in F_A

and ΔH_{r3} vary with the temperature T and the Lie bracket $[b_1(x), b_2(x)]$ has the form $[0\ 0\ 0\ 0\ *]^T$ where '$*$' denotes a nonzero term. Thus, from Theorem 6.2, an ϵ-dependent coordinate change, singular at $\epsilon = 0$, is required to obtain a standard form representation. Moreover, it can be verified that for $\bar{p} = 1$, condition (ii) of Proposition 6.1 is not satisfied for *any* choice of the one-dimensional involutive distribution $\bar{B}(x) = span\{b_j(x)\}$ for $j = 1$ or $j = 2$, i.e., $L_{b_1} k_2(x) \not\equiv 0$, and $L_{b_2} k_1(x) \not\equiv 0$. Thus, both constraints $k_1(x) = 0$ and $k_2(x) = 0$ have to be scaled by the factor $1/\epsilon$ in the coordinate change

$$
\begin{bmatrix} \zeta_1 \\ \zeta_2 \\ \zeta_3 \\ \eta_1 \\ \eta_2 \end{bmatrix} = T(x, \epsilon) = \begin{bmatrix} x_1 + x_2 \\ x_3 + x_4 \\ x_5 \\ \dfrac{1}{\epsilon} e^{\frac{E_1}{R}(\frac{1}{T^0} - \frac{1}{x_5})}(x_1 - \dfrac{x_2}{\kappa_1}) \\ \dfrac{1}{\epsilon} e^{\frac{E_3}{R}(\frac{1}{T^0} - \frac{1}{x_5})} x_3 \end{bmatrix}
$$

to obtain a standard form representation of the form in Eq.6.28. The detailed representation is omitted for brevity.

For the resulting system in the standard form, the two-dimensional fast subsystem has an equilibrium point at:

$$
\eta_{1s} = \alpha_1(\zeta) + \alpha_2(\zeta)u_1 + \alpha_3(\zeta)u_2
$$

$$
\eta_{2s} = \frac{k_2 \kappa_1 \zeta_1}{\kappa_2(\kappa_1 + 1)}
$$

119

where

$$\alpha_1 = \left(\frac{k_2\zeta_1}{\gamma(\zeta)(1+\kappa_1)}\right)\left\{1 - \frac{\kappa_1\zeta_1\Delta H_{r1}}{(1+\kappa_1)R\zeta_3^2\rho c_p}(\Delta H_{r2} + \Delta H_{r3})\right\}$$

$$\alpha_2 = \frac{1}{\gamma(\zeta)}\left\{\frac{C_{Ai}}{V} + \frac{(T_i - \zeta_3)\zeta_1\Delta H_{r1}}{(1+\kappa_1)R\zeta_3^2 V}\right\}$$

$$\alpha_3 = -\frac{\zeta_1\Delta H_{r1}}{\gamma(\zeta)(1+\kappa_1)R\zeta_3^2\rho V c_p}$$

$$\gamma(\zeta) = 1 + \frac{1}{\kappa_1} + \frac{\zeta_1(\Delta H_{r1})^2}{(1+\kappa_1)R\zeta_3^2\rho c_p}$$

Defining the deviation variables $\bar{\eta}_1 = \eta_1 - \eta_{1s}$ and $\bar{\eta}_2 = \eta_2 - \eta_{2s}$, the following representation of the fast subsystem is obtained:

$$\frac{d\bar{\eta}_1}{d\tau} = -\beta_1(\zeta)\,\bar{\eta}_1 - \beta_2(\zeta)\,\bar{\eta}_2$$

$$\frac{d\bar{\eta}_2}{d\tau} = -\kappa_2\,\bar{\eta}_2 \tag{7.8}$$

where:

$$\beta_1(\zeta) = \left[\frac{\kappa_1 + 1}{\kappa_1} + \frac{\zeta_1(\Delta H_{r1})^2}{(1+\kappa_1)\rho c_p R\zeta_3^2}\right] > 0$$

$$\beta_2(\zeta) = \frac{\zeta_1\kappa_2\Delta H_{r1}\Delta H_{r3}}{(1+\kappa_1)\rho c_p R\zeta_3^2}$$

Clearly, the fast subsystem is exponentially stable at $\bar{\eta}_i = 0$, $i = 1, 2$, uniformly in ζ. On the other hand, the three-dimensional slow subsystem has the following representation:

$$\dot{\zeta}_1 = -\frac{k_2\kappa_1}{1+\kappa_1}\zeta_1 + \frac{C_{Ai} - \zeta_1}{V}u_1$$

$$\dot{\zeta}_2 = \frac{k_2\kappa_1}{1+\kappa_1}\zeta_1 - \frac{\zeta_2}{V}u_1$$

$$\dot{\zeta}_3 = -\frac{k_2\kappa_1\zeta_1\Delta H_{r2}}{(1+\kappa_1)\rho c_p} + \left(\frac{T_i - \zeta_3}{V}\right)u_1 - \frac{u_2}{\rho V c_p} - \frac{\Delta H_{r1}}{\rho c_p}\eta_{1s} - \frac{\kappa_2\Delta H_{r3}}{\rho c_p}\eta_{2s}$$

$$y_{1s} = \frac{\kappa_1}{1+\kappa_1}\zeta_1$$

$$y_{2s} = \zeta_3 \tag{7.9}$$

For the above slow subsystem, the relative order of the two outputs y_{1s}, y_{2s} are $r_1 = 1$, $r_2 = 1$, the characteristic matrix is nonsingular, and the one-dimensional zero

120

dynamics is locally exponentially stable. Thus, an input/output linearizing controller was designed on the basis of this slow subsystem, to induce the following decoupled first-order response in the closed-loop slow system:

$$y_{is} + \beta_i \, \dot{y}_{is} = y_{isp}, \quad i = 1, 2$$

with $\beta_1 = 5$ min and $\beta_2 = 8$ min. It was verified that the value of $\epsilon = 0.02$ ensures that the output of the closed-loop system satisfies $y_i(t) = y_{is}(t) + O(\epsilon)$, $t \geq 0$.

Figure 7.5 shows the closed-loop profiles of the controlled outputs and manipulated inputs, for a 10% increase in the setpoint for the first output y_{1sp}. The process at $t = 0$ was at the nominal steady state, which guarantees that $k_1(x) = O(\epsilon)$ and $k_2(x) = O(\epsilon)$. The profiles clearly show an excellent tracking performance with stability. The application study clearly demonstrates the effectiveness of the controller designed on the basis of the slow subsystem of the two-time-scale process that exhibits a slightly nonminimum phase behavior.

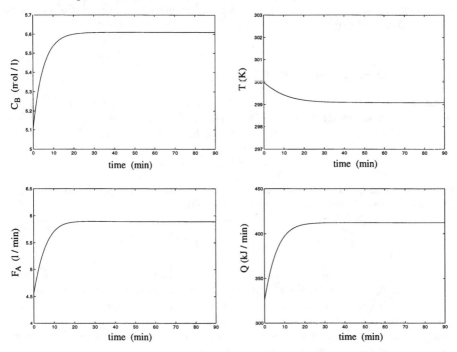

Figure 7.5: Closed-loop input and output profiles under the controller based on the reduced-order slow subsystem, for a 10% increase in the setpoint y_{1sp}.

Remark 7.1: In the above CSTR example, the slightly nonminimum phase behavior, in particular, the inverse response in the fast boundary layer (see Figure 7.4), arises specifically due to a competition between the fast reaction $A \rightleftharpoons B$ and the slow reaction $B \to C$ in series, in the production of the intermediate component B which is the desired product. Such slightly nonminimum phase behavior arising due to a competition between fast and slow phenomena may occur quite often in fast-rate chemical processes, e.g., multi-phase reactors with a fast inter-phase mass transfer followed by a slow reaction in the reaction phase.

Remark 7.2: Note that for the CSTR example with multiple reactions, the controller synthesized on the basis of the slow subsystem does not require the specific values of the kinetic rate constants of the fast reactions, which may be difficult to obtain in practice. Only the kinetics of the slow reaction that essentially governs the process dynamics, and the thermodynamic equilibrium constant for the fast reversible reaction are required in the controller design. Similarly, in processes with fast mass transfer, heat transfer, etc., a knowledge of the specific mass/heat transfer rate coefficients is not required, rather the thermodynamic phase/thermal equilibrium relations are used in the controller design.

7.3 Two-Phase Reactor

Consider the two-phase reactor in Figure 2.2, where a gaseous reactant A and a liquid reactant B are fed to the vapor and liquid phases, respectively, and the product C is formed in the liquid phase through the exothermic reaction $A + B \to C$ at a rate:

$$R_C = k_o \exp(-E_a/RT) M_l \rho x_A x_B$$

Reactant A and product C are volatile, and their saturation vapor pressures P_A^s, P_C^s at a temperature T are given by the Antoine relations:

$$P_A^s = \exp\left(28.0 - \frac{4000}{T - 34}\right), \qquad P_C^s = \exp\left(28.5 - \frac{5000}{T + 70}\right) \qquad (7.10)$$

whereas reactant B is relatively nonvolatile. The liquid stream is withdrawn from the reactor at a flow rate F_l, while the product vapor stream is withdrawn at a rate F_v. The process is open-loop unstable at the nominal steady state (see Table 7.3), and it is desired to control the product vapor composition y_A and reactor temperature T, using the product vapor flow rate F_v and heat output Q as the manipulated inputs (for a detailed description of the process, see section 2.3).

A detailed dynamic model of the process is given by a DAE system (Eq.2.8), which is in the form of Eq.3.1 with the differential variables $x = [M_v \ y_A \ M_l \ x_A \ x_B \ T]^T$, algebraic variables $z = [N_A \ N_C \ P]^T$, manipulated inputs $u = [F_v \ Q]^T$, outputs

$y_1 = y_A$, $y_2 = T$, and

$$f(x) = \begin{bmatrix} F_{A_o} \\ \dfrac{F_{A_o}(1 - x_2)}{x_1} \\ F_{B_o} - F_L - R_C \\ -\left(\dfrac{F_{B_o}x_4 + R_C(1 - x_4)}{x_3}\right) \\ \left(\dfrac{(F_{B_o} - R_C)(1 - x_5)}{x_3}\right) \\ \left(\dfrac{1}{x_1 + x_3}\right)[F_{A_o}(T_{A_o} - x_6) + F_{B_o}(T_{B_o} - x_6) + R_C(x_6 - T_o - \dfrac{\Delta H_R^o}{c_p})] \end{bmatrix}$$

Table 7.3: Reactor parameters and variables and their nominal values

Variable	Description	Nominal value
c_p	molar heat capacity ($J/mole\ K$)	80.0
E_a	activation energy ($kJ/mole$)	100.0
F_{Ao}	inlet vapor stream molar flow rate ($moles/s$)	85.0
F_{Bo}	inlet liquid stream molar flow rate ($moles/s$)	80.0
F_L	outlet liquid stream molar flow rate ($moles/s$)	80.0
F_V	outlet vapor stream molar flow rate ($moles/s$)	57.93
k_o	pre-exponential factor ($m^3/moles\ s$)	1.0e+10
M_l	liquid phase molar holdup ($kmoles$)	4.08
M_v	vapor phase molar holdup ($kmoles$)	0.514
Q	heat output (kW)	890.09
R	universal gas constant ($J/mole\ K$)	8.314
T	reactor temperature (K)	340.0
T_{Ao}	inlet vapor stream temperature (K)	298.0
T_{Bo}	inlet liquid stream temperature (K)	298.0
V_T	reactor volume (m^3)	1.5
x_A	mole fraction of species A in liquid phase	0.29
x_B	mole fraction of species B in liquid phase	0.662
y_A	mole fraction of species A in vapor phase	0.60
ΔH_R^o	heat of reaction ($kJ/mole$)	−40.0
ΔH^v	latent heat of vaporization ($kJ/mole$)	10.0
ρ	liquid phase molar density ($kmoles/m^3$)	8.0

$$b(x) = \begin{bmatrix} -1 & 1 & 0 \\ -(\dfrac{1-x_2}{x_1}) & -(\dfrac{x_2}{x_1}) & 0 \\ 1 & -1 & 0 \\ (\dfrac{1-x_4}{x_3}) & (\dfrac{x_4}{x_3}) & 0 \\ -(\dfrac{x_5}{x_3}) & (\dfrac{x_5}{x_3}) & 0 \\ (\dfrac{\Delta H^v}{(x_1+x_3)c_p}) & -(\dfrac{\Delta H^v}{(x_1+x_3)c_p}) & 0 \end{bmatrix}, \quad g(x) = \begin{bmatrix} -1 & 0 \\ 0 & 0 \\ 0 & 0 \\ 0 & 0 \\ 0 & 0 \\ 0 & (\dfrac{1}{(x_1+x_3)c_p}) \end{bmatrix}$$

$$k(x) = \begin{bmatrix} -x_4 P_A^s \\ -(1-x_4-x_5)P_C^s \\ -Rx_1x_6 \end{bmatrix}, \quad l(x) = \begin{bmatrix} 0 & 0 & x_2 \\ 0 & 0 & (1-x_2) \\ 0 & 0 & \dfrac{(V_T\rho - x_3)}{\rho} \end{bmatrix}, \tag{7.11}$$

$$c(x) = [0]_{3\times 2}, \quad h_1(x) = x_2, \quad h_2(x) = x_6$$

Clearly, the matrix $l(x)$ is singular and the algebraic equations can not be solved for the inter-phase mole transfer rates N_A and N_C, i.e., the DAE model has a high index. In the next section, a state-space realization of the DAE model is derived following the proposed algorithm.

7.3.1 State-space realization

Iteration 1:

Consider the original algebraic equations:

$$0 = k(x) + l(x)z$$

where $k(x), l(x)$ are given in Eq.7.11 and

$$\text{rank } l(x) = p_1 = 1$$

Moreover, the rank condition in Eq.3.15 is trivially satisfied, since $c(x) \equiv 0$.
Step 1. The algebraic equations are premultiplied by the matrix:

$$E^1(x) = \begin{bmatrix} 0 & 0 & 1 \\ 1 & 0 & -(\dfrac{x_2\rho}{V_T\rho - x_3}) \\ 0 & 1 & -(\dfrac{(1-x_2)\rho}{V_T\rho - x_3}) \end{bmatrix}$$

to obtain:

$$0 = - \begin{bmatrix} Rx_1x_6 \\ P_A^s x_4 - (\dfrac{\rho Rx_1x_2x_6}{V_T\rho - x_3}) \\ P_C^s(1 - x_4 - x_5) - (\dfrac{\rho Rx_1(1 - x_2)x_6}{V_T\rho - x_3}) \end{bmatrix} + \begin{bmatrix} 0 & 0 & (\dfrac{V_T\rho - x_3}{\rho}) \\ 0 & 0 & 0 \\ 0 & 0 & 0 \end{bmatrix} \begin{bmatrix} z_1 \\ z_2 \\ z_3 \end{bmatrix} \qquad (7.12)$$

Clearly, the last two equations denote constraints in the differential variables x:

$$0 = \begin{bmatrix} \mathbf{k}_1^1(x) \\ \mathbf{k}_2^1(x) \end{bmatrix} = \begin{bmatrix} \exp(28.0 - \dfrac{4000}{x_6 - 34})x_4 - (\dfrac{\rho Rx_1x_2x_6}{V_T\rho - x_3}) \\ \exp(28.5 - \dfrac{5000}{x_6 + 70})(1 - x_4 - x_5) - (\dfrac{\rho Rx_1(1 - x_2)x_6}{V_T\rho - x_3}) \end{bmatrix} \qquad (7.13)$$

Step 2. The constraints in Eq.7.13 are differentiated once, to obtain the following set of algebraic equations:

$$0 = \begin{bmatrix} Rx_1x_6 \\ \widetilde{k}_1^2(x) \\ \widetilde{k}_2^2(x) \end{bmatrix} + \begin{bmatrix} 0 & 0 & (\dfrac{V_T\rho - x_3}{\rho}) \\ \widetilde{l}_{11}^2(x) & \widetilde{l}_{12}^2(x) & 0 \\ \widetilde{l}_{21}^2(x) & \widetilde{l}_{22}^2(x) & 0 \end{bmatrix} \begin{bmatrix} z_1 \\ z_2 \\ z_3 \end{bmatrix} + \begin{bmatrix} 0 & 0 \\ \widetilde{c}_{11}^2(x) & \widetilde{c}_{12}^2(x) \\ \widetilde{c}_{21}^2(x) & \widetilde{c}_{22}^2(x) \end{bmatrix} \begin{bmatrix} u_1 \\ u_2 \end{bmatrix} \qquad (7.14)$$

A detailed description of the individual terms in Eq.7.14 is omitted here for brevity.
Step 3. The rank of the matrix:

$$\begin{bmatrix} 0 & 0 & (\dfrac{V_T\rho - x_3}{\rho}) \\ \widetilde{l}_{11}^2(x) & \widetilde{l}_{12}^2(x) & 0 \\ \widetilde{l}_{21}^2(x) & \widetilde{l}_{22}^2(x) & 0 \end{bmatrix}$$

is $p_2 = 3 = p$. Thus, the algorithm converged after $s = 1$ iteration implying that the DAE model has an index $\nu_d = 2$. Moreover, the DAE model is also regular, since the two constraints in x (Eq.7.13) are independent of the manipulated inputs u.

The final set of algebraic equations in Eq.7.14 can be solved for the algebraic variables, in particular for $z_1 = N_A$ and $z_2 = N_C$:

$$\begin{bmatrix} z_1 \\ z_2 \end{bmatrix} = - \begin{bmatrix} \widetilde{l}_{11}^2(x) & \widetilde{l}_{12}^2(x) \\ \widetilde{l}_{21}^2(x) & \widetilde{l}_{22}^2(x) \end{bmatrix}^{-1} \left\{ \begin{bmatrix} \widetilde{k}_1^2(x) \\ \widetilde{k}_2^2(x) \end{bmatrix} + \begin{bmatrix} \widetilde{c}_{11}^2(x) & \widetilde{c}_{12}^2(x) \\ \widetilde{c}_{21}^2(x) & \widetilde{c}_{22}^2(x) \end{bmatrix} \begin{bmatrix} u_1 \\ u_2 \end{bmatrix} \right\}$$

$$= \alpha(x) + \beta(x)u \qquad (7.15)$$

Substituting the above solution for z_1, z_2 in the differential equations of the DAE model (note that the differential equations do not involve $z_3 = P$, i.e., the third column of $b(x)$ is identically zero), yields a state-space realization of dimension $n = 6$:

$$\dot{x} = f(x) + \bar{b}(x)\alpha(x) + (g(x) + \bar{b}(x)\beta(x))u$$
$$y_1 = x_2$$
$$y_2 = x_6 \tag{7.16}$$

on the constrained state space $\mathcal{M} = \{x \in \mathbb{R}^6 \; : \; \mathbf{k}_1^1(x) = 0, \mathbf{k}_2^1(x) = 0\}$ ($\bar{b}(x)$ denotes the matrix comprised of the first two columns of $b(x)$ in Eq.7.11).

Following the result of Proposition 3.3, the two constraints in Eq.7.13 can be used as a part of coordinate change to obtain a minimal-order state-space realization (of dimension $\kappa = 4$) of the DAE model. More specifically, note that for the two constraints in Eq.7.13, the Jacobian:

$$\begin{bmatrix} \dfrac{\partial \mathbf{k}_1^1(x)}{\partial x_1} & \dfrac{\partial \mathbf{k}_1^1(x)}{\partial x_2} \\[2mm] \dfrac{\partial \mathbf{k}_2^1(x)}{\partial x_1} & \dfrac{\partial \mathbf{k}_2^1(x)}{\partial x_2} \end{bmatrix} = -\frac{\rho R x_6}{(V_T \rho - x_3)} \begin{bmatrix} x_2 & x_1 \\ 1 - x_2 & -x_1 \end{bmatrix} \tag{7.17}$$

is nonsingular. Thus, the choice of $\phi_1(x) = x_3$, $\phi_2(x) = x_4$, $\phi_3(x) = x_5$, $\phi_4(x) = x_6$ yields a valid coordinate change:

$$\zeta = \begin{bmatrix} \zeta_1^{(0)} \\ \zeta_2^{(0)} \\ \zeta_3^{(0)} \\ \zeta_4^{(0)} \\ \zeta_1^{(1)} \\ \zeta_2^{(1)} \end{bmatrix} = T(x) = \begin{bmatrix} x_3 \\ x_4 \\ x_5 \\ x_6 \\ \mathbf{k}_1^1(x) \\ \mathbf{k}_2^1(x) \end{bmatrix} \tag{7.18}$$

In these coordinates ζ, the states $\zeta_1^{(1)}$, $\zeta_2^{(1)}$ are identically zero on the state space $\mathcal{M}$, and the state-space realization of dimension four obtained by eliminating these zero states, has the following form:

$$\dot{\zeta}_1^{(0)} = f_3(x) + \bar{b}_3(x)\alpha(x) + (g_3(x) + \bar{b}_3(x)\beta(x))u$$
$$\dot{\zeta}_2^{(0)} = f_4(x) + \bar{b}_4(x)\alpha(x) + (g_4(x) + \bar{b}_4(x)\beta(x))u$$
$$\dot{\zeta}_3^{(0)} = f_5(x) + \bar{b}_5(x)\alpha(x) + (g_5(x) + \bar{b}_5(x)\beta(x))u$$
$$\dot{\zeta}_4^{(0)} = f_6(x) + \bar{b}_6(x)\alpha(x) + (g_6(x) + \bar{b}_6(x)\beta(x))u$$
$$y_1 = \frac{P_A^s(\zeta_4^{(0)})\zeta_2^{(0)}}{P_A^s(\zeta_4^{(0)})\zeta_2^{(0)} + P_C^s(\zeta_4^{(0)})(1 - \zeta_2^{(0)} - \zeta_3^{(0)})}$$
$$y_2 = \zeta_4^{(0)} \tag{7.19}$$

126

where $f_i(x)$ denotes the ith component of $f(x)$, and $\bar{b}_i(x), g_i(x)$ denote the ith rows of the corresponding matrices evaluated at $x = T^{-1}(\zeta^{(0)}, 0)$ given by:

$$
x = \begin{bmatrix} (\frac{V_T\rho - \zeta_1^{(0)}}{\rho R \zeta_4^{(0)}})(P_A^s(\zeta_4^{(0)})\zeta_2^{(0)} + P_C^s(\zeta_4^{(0)})(1 - \zeta_2^{(0)} - \zeta_3^{(0)})) \\ \dfrac{P_A^s(\zeta_4^{(0)})\zeta_2^{(0)}}{P_A^s(\zeta_4^{(0)})\zeta_2^{(0)} + P_C^s(\zeta_4^{(0)})(1 - \zeta_2^{(0)} - \zeta_3^{(0)})} \\ \zeta_1^{(0)} \\ \zeta_2^{(0)} \\ \zeta_3^{(0)} \\ \zeta_4^{(0)} \end{bmatrix}
$$

Remark 7.3: The differential equations for $x_3, \dots, x_6$ in the DAE model of Eq.2.8 involve the algebraic variables $z_1 = N_A$ and $z_2 = N_C$, and thus, the state-space realization in Eq.7.19 requires the solution for these algebraic variables (Eq.7.15). It is possible to obtain a state-space realization of the DAE model without evaluating the solution for z_1 or z_2, in appropriate coordinates. More specifically, it can be verified that the distribution $B(x) = \text{span} \{b_1(x), b_2(x)\}$ is involutive, and thus, from Frobenius theorem (see, e.g., [64]), the states $\zeta_i^{(0)}$, or equivalently the functions $\phi_i(x)$, in the coordinate change of Eq.7.18 can be chosen such that $L_{b_j} \phi_i(x) \equiv 0$, $i = 1, \dots 4$, $j = 1, 2$. One such choice of coordinates is as follows:

$$
\zeta = \begin{bmatrix} \zeta_1^{(0)} \\ \zeta_2^{(0)} \\ \zeta_3^{(0)} \\ \zeta_4^{(0)} \\ \zeta_1^{(1)} \\ \zeta_2^{(1)} \end{bmatrix} = T(x) = \begin{bmatrix} M \\ M_A \\ M_B \\ H \\ k_1^1(x) \\ k_2^1(x) \end{bmatrix} \tag{7.20}
$$

where $M = x_1 + x_2$ is the total (in both phases) molar holdup, $M_A = x_1 x_2 + x_3 x_4$ is the total molar holdup of component A, $M_B = x_3 x_5$ is the total molar holdup of component B and $H = (x_1 + x_3)c_p(x_6 - T_o) + x_1 \Delta H^v$ is the total enthalpy. The corresponding state-space realization has the form:

$$
\begin{aligned} \dot{\zeta}^{(0)} &= \widehat{f}(\zeta^{(0)}) + \widehat{g}(\zeta^{(0)})u \\ 0 &= \widehat{h}(y, \zeta^{(0)}) \end{aligned} \tag{7.21}
$$

where $y = [y_1 \cdots y_m]^T$ is the output vector, and the vector fields $\widehat{f}, \widehat{h}$ and the matrix $\widehat{g}$ depend only on $f(x), g(x), h_i(x)$ (Eq.7.11)) and $\phi_i(x)$. Note that the derivation of the above state-space realization essentially amounts to the "modeling" approach for phase-equilibrium processes proposed in [116], where the dynamic balances are written in terms of extensive variables like total holdup and enthalpy, instead of the intensive variables like composition and temperature. It should be mentioned, however, that in the state-space realization of Eq.7.21, $\hat{f}$ and $\hat{g}$ are implicit functions of the extensive state variables $\zeta^{(0)}$, since the reaction rates, phase equilibrium relations, etc., are functions of the intensive variables like T, x_A, x_B, which are related to $\zeta^{(0)}$ through highly nonlinear implicit functions. Furthermore, typically, the outputs y to be controlled are also intensive variables, which can not be expressed as explicit functions of the states $\zeta^{(0)}$. Thus, while the state-space realization in Eq.7.21 can be used for numerical simulation purposes, it is not suitable for feedback controller synthesis.

7.3.2 Controller synthesis

On the basis of the state-space realization in Eq.7.16, or equivalently Eq.7.19, the relative orders r_1 and r_2 of the outputs y_1 and y_2 with respect to the manipulated input vector u are evaluated to be as follows:

$$r_1 = 1; \quad r_2 = 1 \qquad (7.22)$$

and the characteristic matrix is nonsingular. Thus, a closed-loop input/output decoupled response of the following form was requested:

$$y_i + \gamma_{i1}^i \frac{dy_i}{dt} = y_{spi}, \quad i = 1, 2 \qquad (7.23)$$

and the requisite controller was designed through the combination of the state feedback controller of Theorem 3.1 to induce the decoupled input/output response:

$$\beta_{i0}^i y_i + \beta_{i1}^i \frac{dy_i}{dt} = v_i, \quad i = 1, 2 \qquad (7.24)$$

between the inputs $v = [v_1 \ v_2]^T$ and the outputs y, and a linear controller with integral action of the form in Eq.3.47. The controller was tuned with the following parameters:

$$\beta_{10}^1 = 1.0, \quad \beta_{11}^1 = 500s$$

$$\beta_{10}^1 = 1.0, \quad \beta_{21}^2 = 600s$$

$$\gamma_{11}^1 = 500s, \quad \gamma_{21}^2 = 600s$$

7.3.3 Discussion of controller performance

The set point tracking and disturbance rejection capabilities of the controller were evaluated through simulations, where the minimal-order state-space realization in Eq.7.19 was used for the process simulation. In all the runs, for the process starting at the nominal steady state, a 20% decrease in y_{sp1} (corresponding to an increased product purity y_C), and a 3% increase in y_{sp2} was imposed at $t=2$ min.

The first simulation run was performed in the nominal case, when there are no modeling errors or disturbances, and the closed-loop input and output profiles are shown in Figure 7.6. Clearly, the controller induces the first-order responses in the two outputs, as requested. The second run addressed the same setpoint tracking capabilities of the controller in the presence of a 5% errors in the process parameters ρ and c_p. The corresponding profiles in Figure 7.7 show an initial perturbation in the outputs as a result of these parametric errors in the model, which is asymptotically rejected by the controller. The next two runs, addressed the performance of the controller for the same setpoint changes, in the presence of unmeasured disturbances in the two reactant feeds. Figures 7.8 and 7.9 show the closed-loop profiles for a 5% increase in the feed flow rates F_A and F_B, respectively, at $t=2$ min. The controller rejects the effects of these disturbances on the outputs with some performance degradation as expected. These simulation runs clearly demonstrate that the controller enforces the requested closed-loop response in the nominal process and successfully rejects the effects of unmeasured disturbances and parametric errors in the process model.

The performance of the controller was also compared with that of an analogous controller designed on the basis of the index-one DAE model, obtained under the assumption of negligible vapor holdup (Eq.2.9). The index-one model is easily reduced to a standard ODE model, by explicitly solving for y_A from the algebraic equations. On the basis of the resulting ODE representation, the relative orders were again found to be $r_1 = 1$, $r_2 = 1$. Thus, the controller was designed to induce the same decoupled first-order input/output responses (with same parameters β, γ) in the closed-loop system. The performance of this index-one DAE model-based controller was compared with that of the index-two model-based controller for the same setpoint changes. In all simulation runs, the minimal-order state-space realization of the index-two DAE model was used for the process simulation.

Figure 7.10 shows a comparison of the closed-loop profiles under the two controllers in the nominal case, i.e., with no disturbances or modeling errors. The performance of the index-one model-based controller (dashed curves) is reasonably good in comparison to the index-two model-based controller (solid curves), with slight overshoot and oscillations. However, unlike the index-two model-based controller, the index-one model-based controller fails for a 5% increase in the reactant feed flow rate F_A. Figure 7.11 shows a comparison of the performance of the two controllers for a small (1%) increase in F_A. Clearly, while the performance of the index-two model-based controller is close to the nominal performance, the index-one model controller exhibits a very poor performance with large overshoots and oscillations. This run clearly demonstrates the severe limitation in the performance of the index-one model-based controller in the presence of small disturbances which are always present in a practical process.

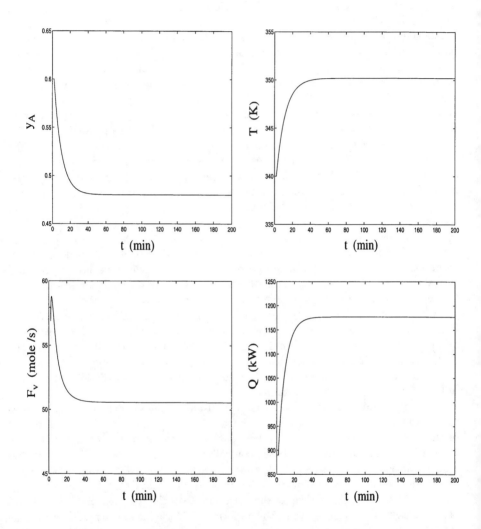

Figure 7.6: Closed-loop input and output profiles in nominal process

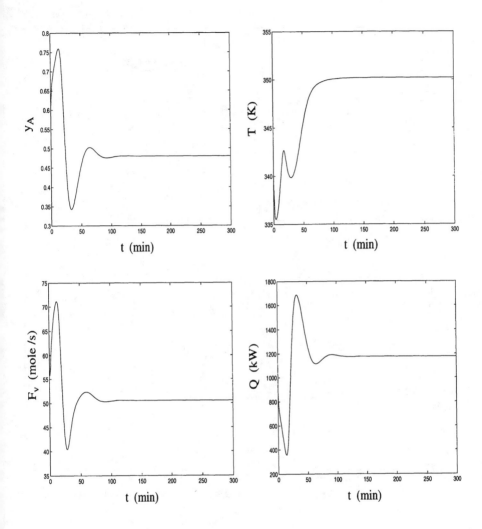

Figure 7.7: Closed-loop input and output profiles in the presence of parametric uncertainties

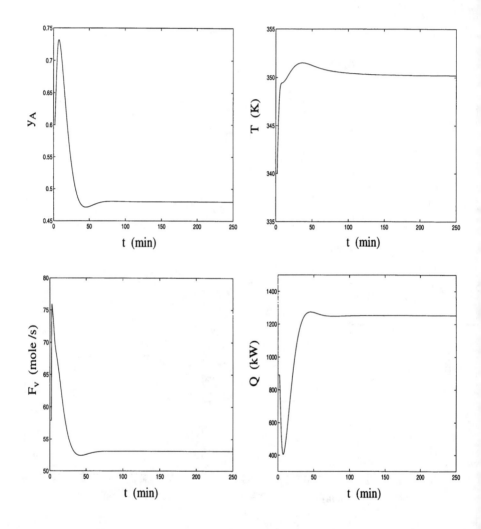

Figure 7.8: Closed-loop input and output profiles in the presence of a disturbance in feed flow rate F_A

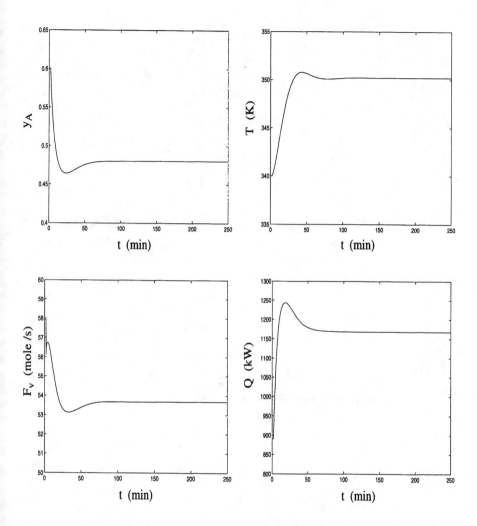

Figure 7.9: Closed-loop input and output profiles in the presence of a disturbance in feed flow rate F_B

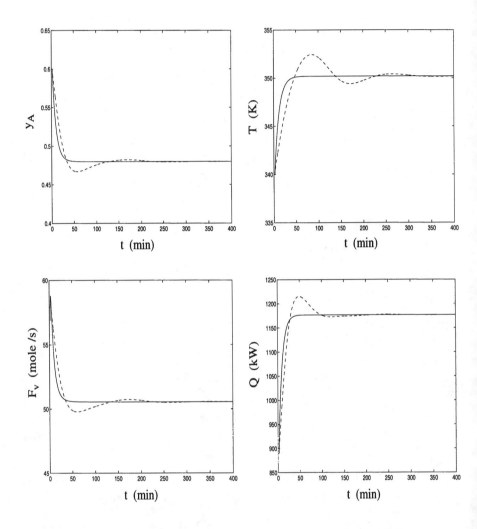

Figure 7.10: Comparison of closed-loop input and output profiles under index-two (solid) and index-one (dashed) model-based controllers, in nominal process

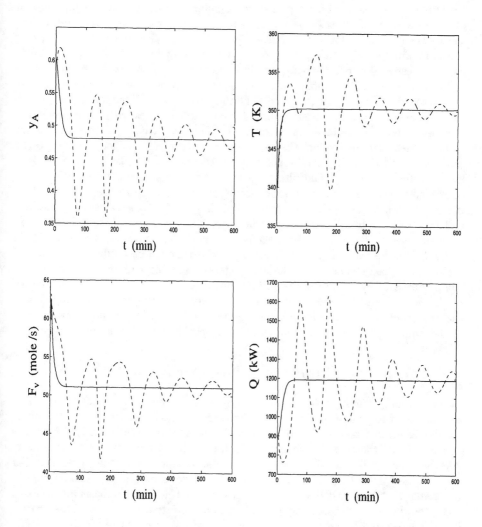

Figure 7.11: Comparison of closed-loop input and output profiles under index-two (solid) and index-one (dashed) model-based controllers, in the presence of a disturbance in feed flow rate F_A

This limitation arises from the inherent inadequacy of the index-one DAE model in describing the dynamic behavior of the process, due to the critical assumption of negligible vapor holdup. Figure 7.12 shows a comparison of the open-loop profiles of the process variables M_l, x_A, x_B and T predicted by the index-two (solid) and index-one (dashed) models, for a 5% increase in the reactant feed flow rate F_A starting at the nominal steady state. Clearly, the two models predict significantly different behavior, especially in the reactant mole fraction x_B. The index-one model, owing to the assumption of negligible vapor holdup, implicitly assumes that *all* the gaseous reactant feed F_A directly enters the liquid (reaction) phase, without any mass transfer (see Eq.2.9). Thus, the model predicts higher reaction rates with a corresponding decrease in both x_A and x_B. In contrast, the index-two model predicts an increase in x_B. This is due to the fact that as the reactor temperature increases due to the exothermic reaction, the relative volatility of the reactant A also increases. Thus, less A diffuses into the liquid phase, i.e., x_A decreases, and consequently less B is consumed in the reaction, i.e., x_B increases (compare also the temperature profiles). It should be mentioned that the two models have exactly the same steady states, and they differ only in their dynamic behavior. The degree of these differences, in general, depends on factors like the relative volatilities of reactants and products, and whether the reaction is exothermic or endothermic. In particular, in the case of endothermic reactions (see [87]), the steady states are stable, and thus, the two models will predict a difference only in the initial transient behavior before the process reaches the (same) new steady state. In contrast, for reactors with exothermic reactions where the operating steady state is unstable, the transient response predicted by the two models could be fundamentally different (see the profiles for x_B in the above example).

7.4 Reactor-Condenser Network

In this section, we address the control of an interconnection of a two-phase reactor and a condenser with liquid recycle, where the reactor pressure dynamics and the interphase mass transfer in the reactor and condenser are fast. While the fast dynamics associated with the mass transfer are stable, the reactor pressure dynamics is unstable; it is, however, easily stabilized at a desired value with a proportional controller. For this process, three DAE models with different degrees of QSS approximations for the stable/stabilized fast dynamics were derived and used as the basis for feedback controller design in [88]. More specifically, a detailed rate-based model of the process is given by an index-one DAE system which is readily reduced to an ODE system. However, the controller designed on the basis of the detailed model leads to instability due to slightly nonminimum phase behavior arising from the fast inter-phase mass transfer. Under the QSS assumption of phase equilibrium for the fast mass transfer, a DAE model of index-two is obtained that is regular. A controller designed on the basis of a state-space realization of this index-two model provides closed-loop stability, but is ill-conditioned. The third model derived under the additional QSS assumption of

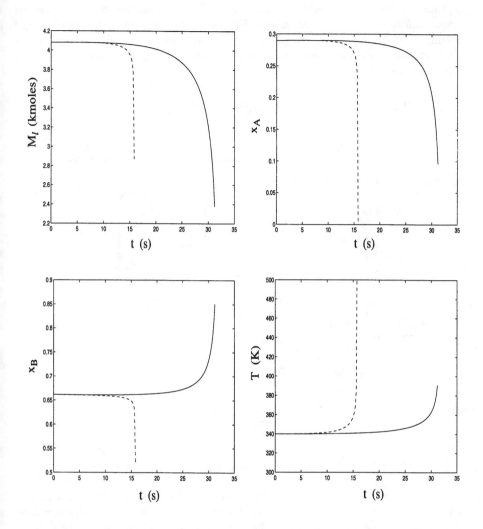

Figure 7.12: Comparison of open-loop profiles for the process states predicted by index-two (solid) and index-one (dashed) models, for a step increase in F_A

constant reactor pressure, is given by an index-three DAE system that is nonregular. The controller designed on the basis of this model through the feedback regularization approach in Chapter 4 provides stability and is not ill-conditioned. For a detailed comparison of the performances of the feedback controllers designed on the basis of the three models, see [88]. In this section, we address the feedforward/feedback control of the process in the presence of measured time-varying disturbances in the reactant feed flow rates.

7.4.1 Process description and modeling

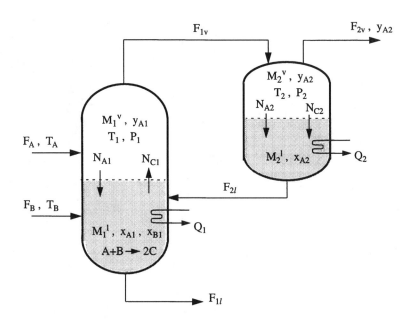

Figure 7.13: A two-phase reactor-condenser network with recycle

Consider the process in Figure 7.13, comprised of a two-phase (vapor/liquid) reactor and a condenser. Reactants A and B are fed to the reactor at molar flow rates F_A, F_B, and temperatures T_A, T_B in the vapor and liquid phases, respectively. Reactant A diffuses into the liquid phase at a rate N_{A1}, where the exothermic reaction:

$$A + B \longrightarrow 2C$$

occurs. Product C diffuses into the vapor phase at a rate N_{C1}, while reactant B is assumed to be nonvolatile. The inter-phase mass transfer resistance is negligible and

138

the reaction rate in the bulk liquid phase is given by the following Arrhenius relation:

$$R_A = k_{10} \exp(\frac{-E_a}{RT_1}) M_1^l \rho x_{A1} x_{B1}$$

where R_A is the rate of consumption of reactant A (or equivalently B) at the temperature T_1, k_{10}, E_a are the pre-exponential factor and activation energy, M_1^l is the liquid molar holdup in the reactor, ρ is the molar density of the liquid, and x_{A1}, x_{B1} are the mole fractions of A and B in the liquid phase, respectively. For simplicity, it is assumed that the molar specific heat capacity c_p, density ρ and latent heat of vaporization ΔH^v are constant and equal for all species and the liquid and vapor phases are ideal mixtures. The liquid stream from the reactor is withdrawn at a constant flow rate F_{1l}, while the vapor flow rate F_{1v} from the reactor to the condenser is governed by the pressure drop $(P_1 - P_2)$. The vapor is cooled in the condenser to a temperature T_2, to enhance the purity of the product C by removing the reactant A into the liquid phase (the relative volatility of C with respect to A increases with decreasing temperature). The liquid phase in the condenser, rich in reactant A, is recycled to the reactor at a flow rate F_{2l}, while the product vapor phase is withdrawn from the condenser at a flow rate F_{2v} and composition y_{A2}.

A detailed dynamic model of the process, that explicitly includes the modeling equations for the vapor holdup in the reactor and condenser, is given by a DAE system. The differential equations include the overall and component mole balances in the liquid and vapor phases and the overall enthalpy balances, in the reactor and the condenser. The algebraic equations on the other hand, include the constitutive relations for the inter-phase mass transfer rates, the ideal gas equations for the vapor holdups and the pressure drop correlation. The overall DAE model for the process is as follows:

$$\dot{M}_1^l = F_B - F_{1l} + F_{2l} + N_{A1} - N_{C1} \tag{7.25}$$

$$\dot{x}_{A1} = (\frac{1}{M_1^l})[-F_B x_{A1} + F_{2l}(x_{A2} - x_{A1}) + N_{A1}(1 - x_{A1}) + N_{C1} x_{A1} - r_A] \tag{7.26}$$

$$\dot{x}_{B1} = (\frac{1}{M_1^l})[F_B(1 - x_{B1}) - F_{2l} x_{B1} - N_{A1} x_{B1} + N_{C1} x_{B1} - r_A] \tag{7.27}$$

$$\dot{M}_1^v = F_A - F_{1v} - N_{A1} + N_{C1} \tag{7.28}$$

$$\dot{y}_{A1} = (\frac{1}{M_1^v})[F_A(1 - y_{A1}) - N_{A1}(1 - y_{A1}) - N_{C1} y_{A1}] \tag{7.29}$$

$$\dot{T}_1 = (\frac{1}{M_1^l + M_1^v})[F_A(T_A - T_1) + F_B(T_B - T_1) + F_{2l}(T_2 - T_1)$$
$$+ (N_{A1} - N_{C1})(\frac{\Delta H^v}{c_p}) - \frac{Q_1}{c_p} + r_A(\frac{-\Delta H_r}{c_p})] \tag{7.30}$$

$$\dot{M}_2^l = N_{A2} + N_{C2} - F_{2l} \tag{7.31}$$

139

$$\dot{x}_{A2} = (\frac{1}{M_2^l}) [N_{A2}(1 - x_{A2}) - N_{C2}x_{A2}] \tag{7.32}$$

$$\dot{M}_2^v = F_{1v} - F_{2v} - N_{A2} - N_{C2} \tag{7.33}$$

$$\dot{y}_{A2} = (\frac{1}{M_2^v}) [F_{1v}(y_{A1} - y_{A2}) - N_{A2}(1 - y_{A2}) + N_{C2}y_{A2}] \tag{7.34}$$

$$\dot{T}_2 = (\frac{1}{M_2^l + M_2^v}) \left[F_{1v}(T_1 - T_2) + (N_{A2} + N_{C2})\frac{\Delta H^v}{c_p} - \frac{Q_2}{c_p} \right] \tag{7.35}$$

$$0 = N_{A1} - k_A a(y_{A1} - y_{A1}^*)\frac{M_1^l}{\rho} \tag{7.36}$$

$$0 = N_{C1} - k_C a(y_{C1}^* - (1 - y_{A1}))\frac{M_1^l}{\rho} \tag{7.37}$$

$$0 = N_{A2} - k_A a(y_{A2} - y_{A2}^*)\frac{M_2^l}{\rho} \tag{7.38}$$

$$0 = N_{C2} - k_C a(1 - y_{A2} - y_{C2}^*)\frac{M_2^l}{\rho} \tag{7.39}$$

$$0 = P_1(V_{1T} - \frac{M_1^l}{\rho}) - M_1^v R T_1 \tag{7.40}$$

$$0 = P_2(V_{2T} - \frac{M_2^l}{\rho}) - M_2^v R T_2 \tag{7.41}$$

$$0 = P_1 - P_2 - \frac{1}{0.09}(F_{1v})^{\frac{7}{4}} \tag{7.42}$$

In the above equations, k_A, k_C denote the overall mass transfer coefficients for A and C, and a denotes the interfacial area per unit liquid holdup volume. Moreover, $y_{A1}^*, y_{C1}^*, y_{A2}^*, y_{C2}^*$ denote compositions of the vapor phase at equilibrium with the liquid phase and are given by the ideal phase equilibrium (Raoult's law) relations:

$$0 = P_1 y_{A1}^* - P_{A1}^s x_{A1} \tag{7.43}$$

$$0 = P_1 y_{C1}^* - P_{C1}^s(1 - x_{A1} - x_{B1}) \tag{7.44}$$

$$0 = P_2 y_{A2}^* - P_{A2}^s x_{A2} \tag{7.45}$$

$$0 = P_2 y_{C2}^* - P_{C2}^s(1 - x_{A2}) \tag{7.46}$$

where P_{Ai}^s, P_{Ci}^s are the saturation vapor pressures of A and C in the reactor $(i = 1)$ and the condenser $(i = 2)$, given by the Antoine relations:

$$P_{Ai}^s = \exp(25.1 - \frac{3400}{T_i + 20})$$

$$P_{Ci}^s = \exp(27.3 - \frac{4100}{T_i + 70})$$

For sufficiently large values of $k_A a$ and $k_C a$, i.e., small mass transfer resistances, the inter-phase mass transfer in the reactor and condenser are fast and stable, and the liquid and vapor phases are close to equilibrium. Furthermore, the dynamics of the pressure P_1 in the reactor is also fast, albeit, unstable. The fast pressure dynamics is stabilized at a desired level P^* by a proportional feedback controller:

$$F_{2v} = F_{2v,nom} - K(P^* - P_1) \qquad (7.47)$$

using the product vapor flow rate F_{2v} from the condenser as the manipulated input. In the above equation, $F_{2v,nom}$ refers to the nominal steady state value of F_{2v} and K is the controller gain. A description and nominal values of the process variables and parameters is included in Table 7.4.

Table 7.4: Description of variables for the reactor-condenser network

Variable	Description	Nominal value
a	mass transfer area per unit liquid holdup volume (m^2/m^3)	1000
c_p	molar heat capacity $(J/mole\ K)$	80.0
E_a	activation energy for the reaction $(kJ/mole)$	110.0
F_A, F_B	inlet molar flow rate of reactant A, B $(moles/s)$	99.84, 52.0
F_{1l}	molar flow rate of liquid stream from reactor $(moles/s)$	10.0
F_{2l}	molar flow rate of liquid recycle from condenser $(moles/s)$	72.19
F_{1v}	molar flow rate of vapor stream from reactor $(moles/s)$	214.03
F_{2v}	molar flow rate of product vapor from condenser $(moles/s)$	141.84
K	proportional gain of pressure controller $(moles/s\ atm)$	10.0
k_{10}	pre-exponential factor $(m^3/mole\ s)$	2.88e+11
k_A, k_C	overall mass transfer coefficient for A, C $(moles/m^2\ s)$	20, 30
M_1^l, M_2^l	liquid molar holdup in reactor, condenser, resp. $(kmoles)$	14.52, 15.0
M_1^v, M_2^v	vapor molar holdup in reactor, condenser, resp. $(kmoles)$	3.75, 3.90
P_1, P_2	pressure in reactor, condenser, resp. (atm)	50.0, 48.69
P^*	setpoint for reactor pressure (atm)	50.0
Q_1, Q_2	heat output from reactor, condenser, resp. (kW)	863.7, 1164.4
T_A, T_B	temperature of reactant feed A, B (K)	315.0, 300.0
T_1, T_2	temperature in reactor, condenser, resp. (K)	330.0, 304.16
V_{1T}, V_{2T}	volume of reactor, condenser, resp. (m^3)	3.0, 3.0
x_{A1}	mole fraction of A in the liquid phase in reactor	0.49
x_{B1}	mole fraction of B in the liquid phase in reactor	0.40
x_{A2}	mole fraction of A in the liquid phase in condenser	0.74
y_{A1}	mole fraction of A in the vapor phase in reactor	0.47
y_{A2}	mole fraction of A in the vapor phase in condenser	0.33
ΔH_r	heat of reaction $(kJ/mole)$	-50.0
ΔH^v	latent heat of vaporization $(kJ/mole)$	10.0
ρ	liquid molar density $(kmoles/m^3)$	15.0

For this process it is desired to control the outputs

$$y_1 = T_1, \quad y_2 = y_{A2}, \quad y_3 = M_2^l$$

141

using the manipulated inputs:

$$u = [\, Q_1 \quad Q_2 \quad F_{2l} \,]^T$$

The detailed DAE model (Eq.7.25-7.47) of the overall process has an index one, and it can be easily reduced to a standard ODE model and used to design an input/output linearizing feedback controller. However, such a controller ignores the inherent time-scale multiplicity in the process and is ill-conditioned, i.e., the control action is highly sensitive to small modeling/measurement errors since the effect of these errors is magnified through the large process parameters, e.g., the large mass transfer coefficients, which appear explicitly in the control law. In fact the closed-loop system is unstable since the process is slightly nonminimum phase; the key controlled output y_{A2} shows an inverse response in the fast boundary layer, for step changes in the inputs F_{2l} and Q_2 [88].

If the fast and stable dynamics associated with the mass transfer is ignored, i.e., the explicit mass transfer correlations (Eq.7.36-7.39) are replaced by the pseudo-steady-state conditions of phase equilibrium:

$$0 = P_1 y_{A1} - P_{A1}^s x_{A1} \tag{7.48}$$

$$0 = P_1(1 - y_{A1}) - P_{C1}^s(1 - x_{A1} - x_{B1}) \tag{7.49}$$

$$0 = P_2 y_{A2} - P_{A2}^s x_{A2} \tag{7.50}$$

$$0 = P_2(1 - y_{A2}) - P_{C2}^s(1 - x_{A2}) \tag{7.51}$$

while the fast reactor pressure dynamics governed by the control equation of Eq.7.47 is retained, the resulting DAE system has an index two and is regular. A feedback controller can be designed on the basis of this index-two model following the approach in Chapter 3. While the closed-loop system under this controller is stable, the controller is still ill-conditioned, implying the need to ignore the stabilized fast dynamics of the reactor pressure P_1 as well, to obtain well-conditioned controllers.

A DAE model describing only the slow dynamics of the process is obtained by ignoring the fast and stable modes associated with the inter-phase mass transfer *and* the reactor pressure dynamics. More specifically, the explicit correlations for the mass transfer rates in Eq.7.36-7.39 are replaced by the phase equilibrium relations in Eq.7.48-7.51, and the control equation for the reactor pressure P_1 in Eq.7.47 is replaced by the corresponding QSS condition:

$$0 = P_1 - P^* \tag{7.52}$$

The differential variables in the DAE system include $M_1^l, x_{A1}, x_{B1}, M_1^v, y_{A1}, T_1, M_2^l, x_{A2}, M_2^v, y_{A2}, T_2$, while the algebraic variables include $P_1, N_{A1}, N_{C1}, F_{1v}, P_2, N_{A2}, N_{C2}, F_{2v}$. However, the algebraic variable F_{1v} appears in a nonlinear fashion in the pressure drop correlation (Eq.7.42). In view of this fact, the following dynamic extension:

$$\dot{F}_{1v} = \overline{F}_{1v} \tag{7.53}$$

142

is employed, where $\overline{F}_{1v}$ replaces F_{1v} as an algebraic variable and F_{1v} is included in an extended vector of differential variables. The resulting DAE system, with the differential equations in Eq.7.25-7.35,7.53, the algebraic equations in Eq.7.48-7.51,7.40-7.42,7.52, the differential variables:

$$x_1 = M_1^l, \quad x_2 = x_{A1}, \quad x_3 = x_{B1}, \quad x_4 = M_1^v, \quad x_5 = y_{A1}, \quad x_6 = T_1$$
$$x_7 = M_2^l, \quad x_8 = x_{A2}, \quad x_9 = M_2^v, \quad x_{10} = y_{A2}, \quad x_{11} = T_2, \quad x_{12} = F_{1v}$$

and the algebraic variables:

$$z_1 = N_{A1}, \quad z_2 = N_{C1}, \quad z_3 = \overline{F}_{1v}, \quad z_4 = N_{A2}, \quad z_5 = N_{C2}, \quad z_6 = F_{2v}, \quad z_7 = P_1, \quad z_8 = P_2$$

is in the form of Eq.4.1, has an index $\nu_d = 3$, and is nonregular; there is one algebraic constraint in the differential variables involving the manipulated inputs u. Thus, a feedback controller can be designed on the basis of this index-three DAE model for the slow process dynamics, using the regularization approach of Chapter 4. The resulting controller is well-conditioned and yields excellent performance compared to the controllers designed on the basis of the index-one and index-two models (see [88]).

In what follows, we consider the control of the process in the presence of measured time-varying disturbances in the reactant feed flow rates F_A and F_B, i.e., $d = [\, F_A \ F_B \,]^T$. More specifically, we address the design of a feedforward/feedback controller on the basis of the index-three DAE model. In this model, there is one underlying algebraic constraint that explicitly involves the disturbances, and thus, a feedforward/feedback regularizing compensator is required to obtain the desired state-space realization of the regularized system.

7.4.2 Feedforward/feedback regularization and state-space realization

Iteration 1:

Step 1. Consider the algebraic equations for the DAE system (Eq.7.48-7.51,7.40-7.42,7.52). They involve only two algebraic variables $z_7 = P_1$ and $z_8 = P_2$, i.e., $p_1 = 2$. Moreover, since the algebraic equations do not involve any of the manipulated inputs or the disturbances, the augmented matrices have ranks $\nu_{1,1} = \nu_{1,2} = m_1 = p_1$. Premultiply the algebraic equations by the nonsingular matrix:

$$E^1(x) = \begin{bmatrix} 0 & 0 & 0 & 0 & 0 & 0 & 0 & 1 \\ 0 & 0 & 0 & 0 & 0 & e_{26} & 0 & 0 \\ 1 & 0 & 0 & 0 & 0 & 0 & 0 & -y_{A1} \\ 1 & 1 & 0 & 0 & 0 & 0 & 0 & -1 \\ 0 & 0 & 0 & 0 & e_{55} & 0 & 0 & -1 \\ 0 & 0 & 1 & 1 & 0 & e_{66} & 0 & 0 \\ 0 & 0 & 0 & 1 & 0 & e_{76} & 0 & 0 \\ 0 & 0 & 0 & 0 & 0 & e_{86} & 1 & -1 \end{bmatrix}$$

where:

$$e_{26} = (\frac{\rho}{\rho V_{2T} - M_2^l}), \quad e_{53} = (\frac{\rho}{\rho V_{1T} - M_1^l})$$

$$e_{66} = -(\frac{\rho}{\rho V_{2T} - M_2^l}), \quad e_{76} = -(\frac{\rho(1 - y_{A2})}{\rho V_{2T} - M_2^l}), \quad e_{86} = (\frac{\rho}{\rho V_{2T} - M_2^l})$$

to obtain the following algebraic equations that can be solved for P_1 and P_2:

$$0 = P_1 - P^*$$
$$0 = P_2 - \frac{\rho M_2^v R T_2}{\rho V_{2T} - M_2^l} \tag{7.54}$$

and six underlying constraints among the differential variables x:

$$0 = \mathbf{k}_1^1(x) = P^* y_{A1} - P_{A1}^s x_{A1}$$

$$0 = \mathbf{k}_2^1(x) = P^* - P_{A1}^s x_{A1} - P_{C1}^s(1 - x_{A1} - x_{B1})$$

$$0 = \mathbf{k}_3^1(x) = P^* - \frac{\rho M_1^v R T_1}{\rho V_{1T} - M_1^l}$$

$$0 = \mathbf{k}_4^1(x) = \frac{\rho M_2^v R T_2}{\rho V_{2T} - M_2^l} - P_{A2}^s x_{A2} - P_{C2}^s(1 - x_{A2}) \tag{7.55}$$

$$0 = \mathbf{k}_5^1(x) = \frac{\rho M_2^v R T_2(1 - y_{A2})}{\rho V_{2T} - M_2^l} - P_{C2}^s(1 - x_{A2})$$

$$0 = \mathbf{k}_6^1(x) = P^* - \frac{\rho M_2^v R T_2}{\rho V_{2T} - M_2^l} - \frac{1}{0.09}(F_{1v})^{\frac{7}{4}}$$

Step 2. Differentiate the above six constraints in x to obtain the new set of algebraic equations with the following form:

$$
0 = \begin{bmatrix} \tilde{k}_1^1 \\ \tilde{k}_2^1 \\ \tilde{k}_1^2 \\ \tilde{k}_2^2 \\ \tilde{k}_3^2 \\ \tilde{k}_4^2 \\ \tilde{k}_5^2 \\ \tilde{k}_6^2 \end{bmatrix} + \begin{bmatrix} 0 & 0 & 0 & 0 & 0 & 0 & 1 & 0 \\ 0 & 0 & 0 & 0 & 0 & 0 & 0 & 1 \\ \tilde{l}_{11}^2 & \tilde{l}_{12}^2 & 0 & 0 & 0 & 0 & 0 & 0 \\ \tilde{l}_{21}^2 & \tilde{l}_{22}^2 & 0 & 0 & 0 & 0 & 0 & 0 \\ \tilde{l}_{31}^2 & \tilde{l}_{32}^2 & 0 & 0 & 0 & 0 & 0 & 0 \\ 0 & 0 & 0 & \tilde{l}_{44}^2 & \tilde{l}_{45}^2 & \tilde{l}_{46}^2 & 0 & 0 \\ 0 & 0 & 0 & \tilde{l}_{54}^2 & \tilde{l}_{55}^2 & \tilde{l}_{56}^2 & 0 & 0 \\ 0 & 0 & \tilde{l}_{63}^2 & \tilde{l}_{64}^2 & \tilde{l}_{65}^2 & \tilde{l}_{66}^2 & 0 & 0 \end{bmatrix} \begin{bmatrix} N_{A1} \\ N_{C1} \\ \overline{F}_{1v} \\ N_{A2} \\ N_{C2} \\ F_{2v} \\ P_1 \\ P_2 \end{bmatrix} + \begin{bmatrix} 0 & 0 & 0 \\ 0 & 0 & 0 \\ \tilde{c}_{11}^2 & 0 & \tilde{c}_{13}^2 \\ \tilde{c}_{21}^2 & 0 & \tilde{c}_{23}^2 \\ \tilde{c}_{31}^2 & 0 & \tilde{c}_{33}^2 \\ 0 & \tilde{c}_{42}^2 & \tilde{c}_{43}^2 \\ 0 & \tilde{c}_{52}^2 & \tilde{c}_{53}^2 \\ 0 & \tilde{c}_{62}^2 & \tilde{c}_{63}^2 \end{bmatrix} \begin{bmatrix} Q_1 \\ Q_2 \\ F_{2l} \end{bmatrix}
$$

$$
+ \begin{bmatrix} 0 & 0 \\ 0 & 0 \\ \tilde{\beta}_{11}^2 & \tilde{\beta}_{12}^2 \\ \tilde{\beta}_{21}^2 & \tilde{\beta}_{22}^2 \\ \tilde{\beta}_{31}^2 & \tilde{\beta}_{32}^2 \\ 0 & 0 \\ 0 & 0 \\ 0 & 0 \end{bmatrix} \begin{bmatrix} F_A \\ F_B \end{bmatrix} \tag{7.56}
$$

where the nonzero terms are functions of the differential variables x, the specific forms of which are omitted for brevity.

The iterative procedure converges in one iteration with $p_2 = p - 1 = 7$, $m_2 = p = 8$, $\nu_{2,2} = 8$, and the algebraic equations in Eq.7.56. Note that the coefficient matrix for z in Eq.7.56 is structurally singular. Specifically, the third, fourth, and fifth algebraic equations involve only two algebraic variables N_{A1} and N_{C1}, and thus, they impose a hidden constraint in x. Premultiply Eq.7.56 with a nonsingular matrix $E^2(x)$ of the form:

$$
E^2(x) = \begin{bmatrix} 1 & 0 & 0 & 0 & 0 & 0 & 0 & 0 \\ 0 & 1 & 0 & 0 & 0 & 0 & 0 & 0 \\ 0 & 0 & 1 & 0 & 0 & 0 & 0 & 0 \\ 0 & 0 & 0 & 1 & 0 & 0 & 0 & 0 \\ 0 & 0 & 0 & 0 & 0 & 1 & 0 & 0 \\ 0 & 0 & 0 & 0 & 0 & 0 & 1 & 0 \\ 0 & 0 & 0 & 0 & 0 & 0 & 0 & 1 \\ 0 & 0 & e_{83} & e_{84} & 1 & 0 & 0 & 0 \end{bmatrix}
$$

where e_{83}, e_{84} are chosen such that:

$$
\begin{bmatrix} e_{83} & e_{84} \end{bmatrix} \begin{bmatrix} \tilde{l}_{11}^2 & \tilde{l}_{12}^2 \\ \tilde{l}_{21}^2 & \tilde{l}_{22}^2 \end{bmatrix} = - \begin{bmatrix} \tilde{l}_{31}^2 & \tilde{l}_{32}^2 \end{bmatrix}
$$

145

to obtain the final set of algebraic equations with the following form:

$$
0 = \begin{bmatrix} \bar{k}_1^1 \\ \bar{k}_2^1 \\ \tilde{k}_1^2 \\ \tilde{k}_2^2 \\ \tilde{k}_4^2 \\ \tilde{k}_5^2 \\ \tilde{k}_6^2 \\ \hat{k}_1 \end{bmatrix} + \begin{bmatrix} 0 & 0 & 0 & 0 & 0 & 0 & 1 & 0 \\ 0 & 0 & 0 & 0 & 0 & 0 & 0 & 1 \\ \tilde{l}_{11}^2 & \tilde{l}_{12}^2 & 0 & 0 & 0 & 0 & 0 & 0 \\ \tilde{l}_{21}^2 & \tilde{l}_{22}^2 & 0 & 0 & 0 & 0 & 0 & 0 \\ 0 & 0 & 0 & \tilde{l}_{44}^2 & \tilde{l}_{45}^2 & \tilde{l}_{46}^2 & 0 & 0 \\ 0 & 0 & 0 & \tilde{l}_{54}^2 & \tilde{l}_{55}^2 & \tilde{l}_{56}^2 & 0 & 0 \\ 0 & 0 & \tilde{l}_{63}^2 & \tilde{l}_{64}^2 & \tilde{l}_{65}^2 & \tilde{l}_{66}^2 & 0 & 0 \\ 0 & 0 & 0 & 0 & 0 & 0 & 0 & 0 \end{bmatrix} \begin{bmatrix} N_{A1} \\ N_{C1} \\ \overline{F}_{1v} \\ N_{A2} \\ N_{C2} \\ F_{2v} \\ P_1 \\ P_2 \end{bmatrix} + \begin{bmatrix} 0 & 0 & 0 \\ 0 & 0 & 0 \\ \tilde{c}_{11}^2 & 0 & \tilde{c}_{13}^2 \\ \tilde{c}_{21}^2 & 0 & \tilde{c}_{23}^2 \\ 0 & \tilde{c}_{42}^2 & \tilde{c}_{43}^2 \\ 0 & \tilde{c}_{52}^2 & \tilde{c}_{53}^2 \\ 0 & \tilde{c}_{62}^2 & \tilde{c}_{63}^2 \\ \hat{c}_{11} & 0 & \hat{c}_{13} \end{bmatrix} \begin{bmatrix} Q_1 \\ Q_2 \\ F_{2l} \end{bmatrix}
$$

$$
+ \begin{bmatrix} 0 & 0 \\ 0 & 0 \\ \tilde{\beta}_{11}^2 & \tilde{\beta}_{12}^2 \\ \tilde{\beta}_{21}^2 & \tilde{\beta}_{22}^2 \\ 0 & 0 \\ 0 & 0 \\ 0 & 0 \\ \hat{\beta}_{11} & \hat{\beta}_{12} \end{bmatrix} \begin{bmatrix} F_A \\ F_B \end{bmatrix} \tag{7.57}
$$

The last equation in Eq.7.57 denotes a constraint in the differential variables x:

$$
0 = \hat{k}_1(x) + \hat{c}_{11}(x)u_1 + \hat{c}_{13}(x)u_3 + \hat{\beta}_{11}(x)d_1 + \hat{\beta}_{12}(x)d_2 \tag{7.58}
$$

which involves the manipulated inputs $u_1 = Q_1$, $u_3 = F_{2l}$ and both the disturbances $d_1 = F_A$, $d_2 = F_B$. Thus, the algorithm yields an equivalent DAE system given by the differential equations in Eq.7.25-7.35,7.53 and the algebraic equations in Eq.7.57. This DAE system is, clearly, not regular. However, the condition of Lemma 5.2 is satisfied, i.e., the disturbances d_1, d_2 can be eliminated from the constraint in Eq.7.58 through a feedforward/feedback regularizing compensator of Section 5.5.

Consider the algebraic equations (Eq.7.57) for the DAE system obtained from the algorithm. For this system, a choice of the following:

$$
M(x) = \begin{bmatrix} 0 & 0 & 1 \\ 0 & 1 & 0 \\ 1 & 0 & -r \end{bmatrix} \tag{7.59}
$$

with $r(x) = \dfrac{\hat{c}_{11}(x)}{\hat{c}_{13}(x)}$ and

$$
S = [0 \ 0 \ 0 \ 0 \ 0 \ 0 \ 1 \ 0 \ 1 \ 0 \ 0 \ 0] \tag{7.60}
$$

satisfy the rank conditions desired in Eq.5.26 and Eq.5.27, respectively. Given these matrices, the feedforward/dynamic feedback regularizing compensator of Theorem 5.2 takes the following form:

$$
\begin{aligned}
\dot{w} &= v_1 \\
Q_1 &= v_3 \\
Q_2 &= v_2 \\
F_{2l} &= (\widehat{c}_{13})^{-1}\left(-\widehat{k}_1 + M_2^l + M_2^v + w - \widehat{c}_{11}v_3 - \widehat{\beta}_{11}(x)d_1 - \widehat{\beta}_{12}(x)d_2\right)
\end{aligned} \tag{7.61}
$$

where $v = [v_1\ v_2\ v_3]^T$ is the new input vector and w is the state of the compensator.

Consider the DAE system with the differential equations in Eq.7.25-7.35,7.53 and the algebraic equations in Eq.7.57, under the feedforward/dynamic state feedback compensator of Eq.7.61. For the resulting DAE system with the extended vector of differential variables $\bar{x} = [x^T\ w]^T$ and the new inputs v, the constraint in Eq.7.58 becomes:

$$
0 = M_2^l + M_2^v + w
$$

Differentiating this constraint, the following set of algebraic equations is obtained:

$$
0 = \begin{bmatrix} \widetilde{k}_1 \\ \widetilde{k}_2 \\ \widetilde{k}_3 \\ \widetilde{k}_4 \\ \widetilde{k}_5 \\ \widetilde{k}_6 \\ \widetilde{k}_7 \\ \widetilde{k}_8 \end{bmatrix} + \begin{bmatrix} 0 & 0 & 0 & 0 & 0 & 0 & 1 & 0 \\ 0 & 0 & 0 & 0 & 0 & 0 & 0 & 1 \\ \widetilde{l}_{11}^2 & \widetilde{l}_{12}^2 & 0 & 0 & 0 & 0 & 0 & 0 \\ \widetilde{l}_{21}^2 & \widetilde{l}_{22}^2 & 0 & 0 & 0 & 0 & 0 & 0 \\ 0 & 0 & 0 & \widetilde{l}_{44}^2 & \widetilde{l}_{45}^2 & \widetilde{l}_{46}^2 & 0 & 0 \\ 0 & 0 & 0 & \widetilde{l}_{54}^2 & \widetilde{l}_{55}^2 & \widetilde{l}_{56}^2 & 0 & 0 \\ 0 & 0 & \widetilde{l}_{63}^2 & \widetilde{l}_{64}^2 & \widetilde{l}_{65}^2 & \widetilde{l}_{66}^2 & 0 & 0 \\ 0 & 0 & 0 & 0 & 0 & -1 & 0 & 0 \end{bmatrix} \begin{bmatrix} N_{A1} \\ N_{C1} \\ \overline{F}_{1v} \\ N_{A2} \\ N_{C2} \\ F_{2v} \\ P_1 \\ P_2 \end{bmatrix} + \begin{bmatrix} 0 & 0 & 0 \\ 0 & 0 & 0 \\ 0 & 0 & \widetilde{c}_{11}^2 - r\widetilde{c}_{13}^2 \\ 0 & 0 & \widetilde{c}_{21}^2 - r\widetilde{c}_{23}^2 \\ 0 & \widetilde{c}_{42}^2 & -r\widetilde{c}_{43}^2 \\ 0 & \widetilde{c}_{52}^2 & -r\widetilde{c}_{53}^2 \\ 0 & \widetilde{c}_{62}^2 & -r\widetilde{c}_{63}^2 \\ 1 & 0 & 0 \end{bmatrix} \begin{bmatrix} v_1 \\ v_2 \\ v_3 \end{bmatrix}
$$

$$
+ \begin{bmatrix} 0 & 0 \\ 0 & 0 \\ \widetilde{\beta}_{11}^2 & \widetilde{\beta}_{12}^2 \\ \widetilde{\beta}_{21}^2 & \widetilde{\beta}_{22}^2 \\ 0 & 0 \\ 0 & 0 \\ 0 & 0 \\ 0 & 0 \end{bmatrix} \begin{bmatrix} F_A \\ F_B \end{bmatrix}
$$

$$
\tag{7.62}
$$

where the coefficient matrix for the algebraic variables z has full rank p. Thus, the above algebraic equations can be solved for the algebraic variables z in terms of the differential variables $\bar{x}$, the inputs v and the disturbances d. Note that the solution for N_{A1} and N_{C1} involve only the input v_3, while the solution for $N_{A2}, N_{C2}, F_{2v}, \overline{F}_{1v}$

involve all the three inputs v_1, v_2, v_3. Moreover, only the solution for N_{A1} and N_{C1} depend on the disturbances d_1, d_2, while the solutions for the other algebraic variables are independent of the disturbances.

A substitution of the solution for the algebraic variables into the differential equations (Eq.7.25-7.35,7.53) yields the state-space realization of Proposition 5.2. More specifically, in view of the nature of the solutions for the algebraic variables mentioned above, the state-space realization has the following form:

$$\dot{\bar{x}}_1 = \overline{f}_1(\bar{x}) + \overline{g}_{1,3}(\bar{x})v_3 + \overline{\alpha}_{11}(\bar{x})d_1 + \overline{\alpha}_{12}(\bar{x})d_2$$

$$\vdots$$

$$\dot{\bar{x}}_6 = \overline{f}_6(\bar{x}) + \overline{g}_{6,3}(\bar{x})v_3 + \overline{\alpha}_{61}(\bar{x})d_1 + \overline{\alpha}_{62}(\bar{x})d_2$$

$$\dot{\bar{x}}_7 = \overline{f}_7(\bar{x}) + \overline{g}_{7,1}(\bar{x})v_1 + \overline{g}_{7,2}(\bar{x})v_2 + \overline{g}_{7,3}(\bar{x})v_3$$

$$\vdots$$

$$\dot{\bar{x}}_{12} = \overline{f}_{12}(\bar{x}) + \overline{g}_{12,1}(\bar{x})v_1 + \overline{g}_{12,2}(\bar{x})v_2 + \overline{g}_{12,3}(\bar{x})v_3$$

$$\dot{\bar{x}}_{13} = v_1$$

$$y_1 = \bar{x}_6$$

$$y_2 = \bar{x}_{10}$$

$$y_3 = \bar{x}_7 \tag{7.63}$$

where $\bar{x}_{13} = w$, and

$$\bar{x} \in \mathcal{M} = \left\{ (x,w) \in \mathcal{X} \times \mathbb{R} : \begin{array}{c} \mathbf{k}_1^1(x) = 0 \\ \vdots \\ \mathbf{k}_6^1(x) = 0 \\ x_7 + x_9 + w = 0 \end{array} \right\}$$

7.4.3 Feedforward/feedback controller synthesis and control performance

Consider the state-space realization in Eq.7.63 of the feedforward/feedback regularized DAE system. It can be verified that the relative orders of the controlled outputs y_1, y_2 and y_3 with respect to the input vector $v = [v_1 \ v_2 \ v_3]^T$ are as follows:

$$r_1 = 1, \quad r_2 = 1, \quad r_3 = 1,$$

the corresponding characteristic matrix with the following form:

$$C(\bar{x}) = \begin{bmatrix} 0 & 0 & \overline{g}_{6,3}(\bar{x}) \\ \overline{g}_{10,1}(\bar{x}) & \overline{g}_{10,2}(\bar{x}) & \overline{g}_{10,3}(\bar{x}) \\ \overline{g}_{7,1}(\bar{x}) & \overline{g}_{7,2}(\bar{x}) & \overline{g}_{7,3}(\bar{x}) \end{bmatrix}$$

is nonsingular, and the relative orders of the outputs with respect to the disturbance input vector d are as follows:

$$\rho_1 = 1, \quad \rho_2 = 2, \quad \rho_3 = 2.$$

Thus, the necessary and sufficient conditions of Eq.5.19 for the solution of the feedforward/feedback controller synthesis problem are trivially satisfied, and the following first-order decoupled input/output responses were requested in the closed-loop system:

$$y_i + \gamma_{i1}^i \frac{dy_i}{dt} = y_{isp}, \quad i = 1, 2, 3 \tag{7.64}$$

where y_{isp} is the setpoint for the output y_i. The overall feedforward/dynamic state feedback controller has the following form:

$$\dot{w} = v_1$$

$$\begin{bmatrix} v_1 \\ v_2 \\ v_3 \end{bmatrix} = \left\{ \begin{bmatrix} \gamma_{11}^1 & 0 & 0 \\ 0 & \gamma_{21}^2 & 0 \\ 0 & 0 & \gamma_{31}^3 \end{bmatrix}^{-1} C(\bar{x}) \right\} \begin{bmatrix} y_{1sp} - x_6 - \gamma_{11}^1(\overline{f}_6(\bar{x}) + \overline{\alpha}_{61}(\bar{x})d_1 + \overline{\alpha}_{62}(\bar{x})d_2) \\ y_{2sp} - x_{10} - \gamma_{21}^2 \overline{f}_{10}(\bar{x}) \\ y_{3sp} - x_7 - \gamma_{31}^3 \overline{f}_7(\bar{x}) \end{bmatrix}$$

$$Q_1 = v_3$$
$$Q_2 = v_2$$

$$F_{2l} = (\widehat{c}_{13}(x))^{-1} \left(-\widehat{k}_1(x) + M_2^l + M_2^v + w - \widehat{c}_{11}(x)v_3 - \widehat{\beta}_{11}(x)d_1 - \widehat{\beta}_{12}(x)d_2 \right) \tag{7.65}$$

The controller was tuned with the parameters:

$$\gamma_{11}^1 = 900 \ s, \quad \gamma_{21}^2 = 600 \ s, \quad \gamma_{31}^3 = 1200 \ s$$

and its performance was studied through simulations, where the process was simulated by the detailed index-one DAE model.

For the process starting at the nominal steady state, a 10% decrease in the setpoint y_{sp2} and a 1% decrease in the setpoint y_{sp1} were imposed at $t = 10 \, min$, while the following slowly varying disturbances were imposed in the feed flow rates at $t = 20 \, min$:

$$F_A = F_{A,nom}(1 + 0.05 \ sin(\frac{2\pi}{10800}(t - 1200))) \tag{7.66}$$

$$F_B = F_{B,nom}(1 + 0.05 \ sin(\frac{2\pi}{10800}(t - 1200))) \tag{7.67}$$

where the subscript *nom* refers to nominal values. Figure 7.14 shows the closed-loop input and output profiles in the absence of any feedforward compensation for the disturbances, i.e., only the nominal values of F_A and F_B are used in the controller. Clearly, the disturbances have a strong influence on the controlled outputs and the closed-loop dynamics.

Figure 7.15 shows the closed-loop input and output profiles under the feedforward/feedback controller in Eq.7.65, for the same setpoint changes and disturbances.

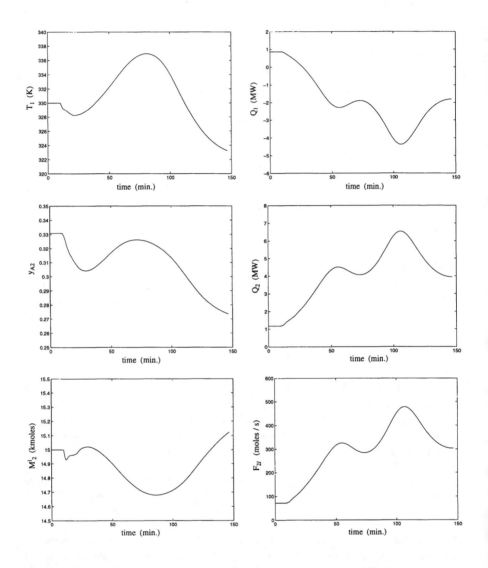

Figure 7.14: Closed-loop profiles for a 1% decrease in y_{sp1} and 10% decrease in y_{sp2}, in the absence of feedforward compensation for disturbances in F_A and F_B.

The controller works well to track the setpoint changes and attenuate the effect of the disturbances. The small oscillations in the outputs correspond to the slight discrepancy between the index-one model used for the process simulation and the index-three model used in controller design; the fast and stable dynamics in the index-one model corresponding to the fast inter-phase mass transfer and reactor pressure dynamics are ignored in the index-three model.

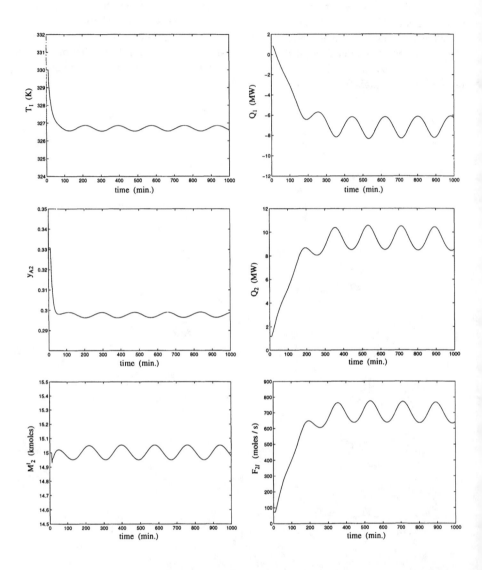

Figure 7.15: Closed-loop profiles for a 1% decrease in y_{sp1} and 10% decrease in y_{sp2}, under the feedforward/feedback controller in the presence of disturbances in F_A and F_B.

Bibliography

[1] A. Ailon, "On the Reduced-Order Causal Observer for Generalized Control Systems," *Int. J. Contr.*, **57**, 1311–1323, 1993.

[2] R. Bachmann, L. Brull, Th. Mrzigold, and U. Pallaske, "On Methods For Reducing The Index Of Differential Algebraic Equations," *Comput. Chem. Engng.*, **14**, 1271–1273, 1990.

[3] J. E. Bailey and D. F. Ollis, *Biochemical Engineering Fundamentals*. McGraw-Hill, Singapore, 1987.

[4] V. B. Bajic, "Nonlinear Functions and Stability of Motions of Implicit Differential Systems," *Int. J. Contr.*, **52**, 1167–1187, 1990.

[5] V. B. Bajic, *Lyapunov's Direct Method in The Analysis of Singular Systems and Networks*. Shades Technical Publications, Hillcrest, Natal, 1992.

[6] D. J. Bender and A. J. Laub, "The Linear-Quadratic Optimal Regulator for Descriptor Systems," *IEEE Trans. Automat. Contr.*, **32**, 672–688, 1987.

[7] J. R. Bowen, A. Acrivos, and A. K. Oppenheim, "Singular Perturbation Refinement to Quasi-Steady State Approximation in Chemical Kinetics," *Chem. Eng. Sci.*, **18**, 177–188, 1963.

[8] K. E. Brenan, *Stability and Convergence of Difference Approximations for Higher Index Differential-Algebraic Systems with Applications in Trajectory Control*, Ph.D. thesis, Dept. of Mathematics, UCLA, Los Angeles, CA, 1983.

[9] K. E. Brenan, S. L. Campbell, and L. R. Petzold, *Numerical Solution of Initial-Value Problems in Differential-Algebraic Equations*, Classics in Applied Mathematics. Society for Industrial and Applied Mathematics, Philadelphia, 1996.

[10] A. Bunse-Gerstner, V. Mehrmann, and N. K. Nichols, "Regularization of Descriptor Systems by Output Feedback," *IEEE Trans. Automat. Contr.*, **39**, 1742–1748, 1994.

[11] G. D. Byrne and P. R. Ponzi, "Differential-Algebraic Systems, Their Applications and Solutions," *Comput. Chem. Engng.*, **12**, 377–382, 1988.

[12] S. L. Campbell, *Singular Systems of Differential Equations*, volume 40 of *Research Notes in Mathematics*. Pitman Pub. Ltd., London, 1980.

[13] S. L. Campbell, "A More Singular Singularly Perturbed Linear System," *IEEE Trans. Automat. Contr.*, **26**, 507–510, 1981.

[14] S. L. Campbell, *Singular Systems of Differential Equations II*, volume 61 of *Research Notes in Mathematics*. Pitman Books Ltd., London, 1982.

[15] S. L. Campbell, "Consistent Initial Conditions for Singular and Nonlinear Systems," *Circ. Sys. Signal Proc.*, **2**, 45–55, 1983.

[16] S. L. Campbell, "A General Form for Solvable Linear Time Varying Singular Systems of Differential Equations," *SIAM J. Math. Anal.*, **18**, 1101–1115, 1987.

[17] S. L. Campbell and C. W. Gear, "The Index of General Nonlinear DAEs," *Numerische Mathematik*, **72**, 173–196, 1995.

[18] S. L. Campbell and E. Griepentrog, "Solvability of General Differential Algebraic Equations," *SIAM J. Sci. Comput.*, **16**, 257–270, 1995.

[19] S. L. Campbell and E. Moore, "Progress on a General Numerical Method for Nonlinear Higher Index DAEs II," *Circ. Sys. Signal Proc.*, **13**, 123–138, 1994.

[20] S. L. Campbell and E. Moore, "Constraint Preserving Integrators for General Nonlinear Higher Index DAEs," *Numerische Mathematik*, **69**, 383–399, 1995.

[21] H. C. Chang and M. Aluko, "Multi-Scale Analysis of Exotic Dynamics in Surface Catalysed Reactions-I," *Chem. Eng. Sci.*, **39**, 37–50, 1984.

[22] X. Chen and M. A. Shayman, "Dynamics and Control of Constrained Nonlinear Systems with Application to Robotics," in *Proc. of Amer. Contr. Conf.*, Chicago, IL, 2962–2966, 1992.

[23] J. H. Chow and P. V. Kokotovic, "A Decomposition of Near Optimum Regulators for Systems with Slow and Fast Modes," *IEEE Trans. Automat. Contr.*, **21**, 701–705, 1976.

[24] J. H. Chow and P. V. Kokotovic, "Two-Time-Scale Feedback Design of a Class of Nonlinear Systems," *IEEE Trans. Automat. Contr.*, **23**, 438–443, 1978.

[25] M. A. Christodoulou and C. Isik, "Feedback Control for Nonlinear Singular Systems," *Int. J. Contr.*, **51**, 487–494, 1990.

[26] M. A. Christodoulou, B. G. Mertzios, and F. L. Lewis, "Simplified Realization Algorithm for Singular Systems," in *Proc. of 26th IEEE Conf. on Dec. and Contr.*, Los Angeles, CA, 1142–1143, 1987.

[27] P. D. Christofides, *Nonlinear Control of Two-Time-Scale and Distributed Parameter Systems*, Ph.D. thesis, Dept. of Chem. Eng. and Mat. Sci., University of Minnesota, Minneapolis, MN, 1996.

[28] P. D. Christofides and P. Daoutidis, "Compensation of Measurable Disturbances in Two-Time-Scale Nonlinear Systems," *Automatica*, **32**, 1553–1573, 1996.

[29] P. D. Christofides and P. Daoutidis, "Feedback Control of Two-Time-Scale Nonlinear Systems," *Int. J. Contr.*, **63**, 965–994, 1996.

[30] Y. Chung and A. W. Westerberg, "A Proposed Numerical Algorithm For Solving Nonlinear Index Problems," *Ind. Eng. Chem. Res.*, **29**, 1234–1239, 1990.

[31] D. Cobb, "Feedback and Pole Placement in Descriptor Variable Systems," *Int. J. Contr.*, **33**, 1135–1146, 1981.

[32] D. Cobb, "On the Solutions of Linear Differential Equations with Singular Coefficients," *J. Diff. Eq.*, **46**, 310–323, 1982.

[33] D. Cobb, "Descriptor Variable Systems and Optimal State Regulation," *IEEE Trans. Automat. Contr.*, **28**, 601–611, 1983.

[34] D. Cobb, "Controllability, Observability, and Duality in Singular Systems," *IEEE Trans. Automat. Contr.*, **29**, 1076–1082, 1984.

[35] J.E. Cuthrell and L. T. Biegler, "On the Optimization of Differential-Algebraic Process Systems," *AIChE J.*, **33**, 1257–1270, 1987.

[36] J.E. Cuthrell and L. T. Biegler, "Simultaneous Optimization and Solution Methods for Batch Reactor Control Profiles," *Comput. Chem. Engng.*, **13**, 49–62, 1989.

[37] L. Dai, *Singular Control Systems*, volume 118 of *Lecture Notes in Control and Information Sciences*. Springer-Verlag, Berlin, Heidelberg, 1989.

[38] P. Daoutidis and C. Kravaris, "Dynamic Compensation of Measurable Disturbances in Nonlinear Multivariable Systems," *Int. J. Contr.*, **58**, 1279–1301, 1993.

[39] P. Daoutidis and C. Kravaris, "Dynamic Output Feedback Control of Multivariable Nonlinear Processes," *Chem. Eng. Sci.*, **49**, 433–447, 1994.

[40] P. Daoutidis and A. Kumar, "Structural Analysis and Output Feedback Control of Nonlinear Multivariable Processes," *AIChE J.*, **40**, 647–669, 1994.

[41] P. Daoutidis, M. Soroush, and C. Kravaris, "Feedforward/Feedback Control of Multivariable Nonlinear Processes," *AIChE J.*, **36**, 1471–1484, 1990.

[42] M. M. Denn, *Process Modeling*. Longam Inc., New York, 1986.

[43] V. Dolezal, "Generalized Solutions of Semistate Equations and Stability," *Circ. Sys. Signal Proc.*, **5**, 391–403, 1986.

[44] N. Fenichel, "Geometric Singular Perturbation Theory for Ordinary Differential Equations," *J. Diff. Equat.*, **31**, 53–98, 1979.

[45] R. Gandikota, I. Karafyllis, and P. Daoutidis, "On the Feedback Control of Incompressible Flows," in *AIChE Ann. Mtg.*, Los Angeles, CA, 1997.

[46] R. Gani and I. T. Cameron, "Modeling for Dynamic Simulation of Chemical Processes: The Index Problem," *Chem. Eng. Sci.*, **47**, 1311–1315, 1992.

[47] F. R. Gantmacher, *The Theory of Matrices*. Chelsea Pub. Co., New York, 1959.

[48] C. W. Gear, "Differential-Algebraic Equation Index Transformations," *SIAM J. Sci. Stat. Comput.*, **9**, 39–47, 1988.

[49] C. W. Gear, "Differential Algebraic Equations, Indices, and Integral Algebraic Equations," *SIAM J. Numer. Anal.*, **27**, 1527–1534, 1990.

[50] C. W. Gear and L. R. Petzold, "ODE Methods For The Solution Of Differential/Algebraic Systems," *SIAM J. Numer. Anal.*, **21**, 716–728, 1984.

[51] M. S. Goodwin, "Exact Pole Assignment with Regularity by Output Feedback in Descriptor Systems – Part 2," *Int. J. Contr.*, **62**, 413–441, 1995.

[52] M. S. Goodwin and L. R. Fletcher, "Exact Pole Assignment with Regularity by Output Feedback in Descriptor Systems – Part 1," *Int. J. Contr.*, **62**, 379–411, 1995.

[53] P. M. Gresho, "Incompressible Fluid Dynamics: Some Fundamental Formulation Issues," *Annu. Rev. Fluid Mech.*, **23**, 413–453, 1991.

[54] P. M. Gresho, "Some Current CFD Issues Relevant to the Incompressible Navier-Stokes Equations," *Computer Methods in App. Mech. and Eng.*, **87**, 201–252, 1991.

[55] P. M. Gresho, S. T. Chan, R. L. Lee, and C. D. Upson, "A Modified Finite Element Method for Solving the Time-Dependent Incompressible Navier-Stokes Equations. Part 1: Theory," *Int. J. Numer. Methods in Fluids*, **4**, 557–598, 1984.

[56] J. Grimm, "Realization and Canonicity for Implicit Systems," *SIAM J. Contr. Optim.*, **26**, 1331–1347, 1988.

[57] E. Hairer, C. Lubich, and M. Roche, *The Numerical Solution of Differential-Algebraic Systems by Runge-Kutta Methods*, Vol. 1409 of *Lecture Notes in Mathematics*. Springer-Verlag, Berlin, Heidelberg, 1989.

[58] G. E. Hayton, P. Fretwell, and A. C. Pugh, "Fundamental Equivalence of Generalized State Space Systems," *IEEE Trans. Automat. Contr.*, **31**, 431–439, 1986.

[59] F. G. Heineken, H. M. Tsuchiya, and R. Aris, "On the Mathematical Status of the Pseudo-Steady State Hypothesis of Biochemical Kinetics," *Mathematical Biosciences*, **1**, 95–121, 1967.

[60] H. Hemami and B. F. Wyman, "Modeling and Control of Constrained Dynamic Systems with Application to Biped Locomotion in the Frontal Plane," *IEEE Trans. Automat. Contr.*, **24**, 526–535, 1979.

[61] R. M. Hirschorn, "Invertibility of Multivariable Nonlinear Control Systems," *IEEE Trans. Automat. Contr.*, **24**, 855–865, 1979.

[62] R. M. Hirschorn, "(A,B)-Invariant Distributions and Disturbance Decoupling of Nonlinear Systems," *SIAM J. Contr. Optim.*, **19**, 1–19, 1981.

[63] M. Hou and P. C. Müller, "Design of a Class of Luenberger Observers for Descriptor Systems," *IEEE Trans. Automat. Contr.*, **40**, 133–136, 1995.

[64] A. Isidori, *Nonlinear Control Systems*. Springer-Verlag, London, third edition, 1995.

[65] A. Isidori, A. J. Krener, C. Gori-Giorgi, and S. Monaco, "Nonlinear Decoupling via Feedback: A Differential Geometric Approach," *IEEE Trans. Automat. Contr.*, **AC-26**, 331–345, 1981.

[66] A. Isidori, S. S. Sastry, P. V. Kokotovic, and C. I. Byrnes, "Singularly Perturbed Zero Dynamics of Nonlinear Systems," *IEEE Trans. Automat. Contr.*, **37**, 1625–1631, 1992.

[67] R. B. Jarvis and C. C. Pantelides, "A Differentiation-Free Algorithm for Solving High-Index DAE Systems," in *AIChE annual meeting 92*, Miami Beach, FL, 1992.

[68] E. Jonckheere, "Variational Calculus for Descriptor Problems," *IEEE Trans. Automat. Contr.*, **33**, 491–495, 1988.

[69] H. K. Khalil, "Feedback Control of Nonstandard Singularly Perturbed Systems," *IEEE Trans. Automat. Contr.*, **34**, 1052–1060, 1989.

[70] H. K. Khalil, *Nonlinear Systems*. Prentice-Hall, Inc., Upper Saddle River, second edition, 1996.

[71] P. V. Kokotovic, "Applications of Singular Perturbation Techniques to Control Problems," *SIAM Rev.*, **26**, 501–550, 1984.

[72] P. V. Kokotovic, A. Bensoussan, and G. Blankenship, *Singular Perturbations and Asymptotic Analysis in Control Systems*, Vol. 90 of *Lecture Notes in Control and Information Sciences*. Springer-Verlag, Berlin, Heidelberg, 1987.

[73] P. V. Kokotovic, O'Malley R. E., and P. Sannuti, "Singular Perturbations and Order Reduction in Control Theory – an Overview," *Automatica*, **12**, 123–132, 1976.

[74] P. V. Kokotovic, H. K. Khalil, and J. O'Reilly, *Singular Perturbations in Control: Analysis and Design*. Academic Press, London, 1986.

[75] H. Krishnan, *Control of Nonlinear Systems with Applications to Constrained Robots and Spacecraft Attitude Stabilization*, Ph.D. thesis, Dept. of Aerospace Eng., The University of Michigan, 1992.

[76] H. Krishnan and N. H. McClamroch, "On Control Systems Described by a Class of Linear Differential-Algebraic: State Realizations and Linear Quadratic Optimal Control," in *Proc. of Amer. Contr. Conf.*, San Diego, CA, 818–823, 1990.

[77] H. Krishnan and N. H. McClamroch, "Tracking in Control Systems Described by Nonlinear Differential-Algebraic Equations with Applications to Constrained Robot Systems," in *Proc. of Amer. Contr. Conf.*, San Francisco, CA, 837, 1993.

[78] H. Krishnan and N. H. McClamroch, "On the Connection Between Nonlinear Differential-Algebraic Equations and Singularly Perturbed Control Systems in Nonstandard Form," *IEEE Trans. Automat. Contr.*, **39**, 1079–1084, 1994.

[79] H. Krishnan and N. H. McClamroch, "Tracking in Nonlinear Differential-Algebraic Control Systems with Applications to Constrained Robot Systems," *Automatica*, **30**, 1885–1897, 1994.

[80] M. Krstic, I. Kanellakopoulos, and P. Kokotovic, *Nonlinear and Adaptive Control Design*. Wiley, New York, 1995.

[81] M. Kuijper and J. M. Schumacher, "Minimality of Descriptor Representations under External Equivalence," *Automatica*, **27**, 985–995, 1991.

[82] A. Kumar, P. D. Christofides, and P. Daoutidis, "Singular Perturbation Modeling of Nonlinear Two-Time-Scale Processes with Nonexplicit Time-Scale Separation," in *Proceedings of 13th IFAC World Congress*, San Francisco, CA, vol. M, 1–6, 1996.

[83] A. Kumar, P. D. Christofides, and P. Daoutidis, "Singular Perturbation Modeling of Nonlinear Processes with Nonexplicit Time-Scale Separation," *Chem. Eng. Sci.*, **53**, 1491–1504, 1998.

[84] A. Kumar and P. Daoutidis, "Control of Nonlinear Differential-Algebraic Process Systems," in *Proc. of Amer. Contr. Conf.*, Baltimore, MD, 330–334, 1994.

[85] A. Kumar and P. Daoutidis, "Control of Nonlinear Differential-Algebraic-Equation Systems with Disturbances," *Ind. Eng. Chem. Res.*, **34**, 2060–2076, 1995.

[86] A. Kumar and P. Daoutidis, "A DAE Framework for Modeling and Control of Reactive Distillation Columns," in *Preprints of 4th IFAC Symp. on Dynamics and Control of Chemical Reactors, Distillation Columns and Batch Processes*, Helsingor, Denmark, 99–104, 1995.

[87] A. Kumar and P. Daoutidis, "Feedback Control of Nonlinear Differential-Algebraic-Equation Systems," *AIChE J.*, **41**, 619–636, 1995.

[88] A. Kumar and P. Daoutidis, "Dynamic Feedback Regularization and Control of Nonlinear Differential-Algebraic-Equation Systems," *AIChE J.*, **42**, 2175–2198, 1996.

[89] A. Kumar and P. Daoutidis, "State-space Realizations of Linear Differential-Algebraic-Equations Systems with Control-dependent State Space," *IEEE Trans. Automat. Contr.*, **41**, 269–274, 1996.

[90] A. Kumar and P. Daoutidis, "High-Index DAE Systems in Modeling and Control of Chemical Processes," in *Preprints of IFAC Conf. on Control of Industrial Systems*, volume 1, Belfort, France, 518–523, 1997.

[91] A. Kumar and P. Daoutidis, "Control of Nonlinear Differential Algebraic Equation Systems: An Overview," in *Nonlinear Model-Based Process Control*, R. Berber and C. Kravaris, editors, NATO-ASI series, Kluwer Academic Publishers, The Netherlands, 1998.

[92] A. Kumar and P. Daoutidis, "Modeling, Analysis, and Control of an Ethylene Glycol Reactive Distillation Column," *AIChE J.*, to appear, 1998.

[93] A. Lefkopoulos and M. A. Stadherr, "Index Analysis of Unsteady-State Chemical Process Systems – II. Strategies for Determining the Overall Flowsheet Index," *Comput. Chem. Engng.*, **17**, 415–430, 1993.

[94] B. Leimkuhler, L. R. Petzold, and C. W. Gear, "Approximation Methods For The Consistent Initialization Of Differential-Algebraic Equations," *SIAM J. Numer. Anal.*, **28**, 205–226, 1991.

[95] F. L. Lewis, "A Survey of Linear Singular Systems," *Circ. Sys. Signal Proc.*, **5**, 3–36, 1986.

[96] J. Lin and Z. Yang, "Existence and Uniqueness of Solutions for Nonlinear Singular (Descriptor) Systems," *Int. J. of Syst. Sci.*, **19**, 2179–2184, 1988.

[97] V. Lovass-Nagy, D. L. Powers, and R. J. Schilling, "On Regularizing Descriptor Systems by Output Feedback," *IEEE Trans. Automat. Contr.*, **39**, 1507–1509, 1994.

[98] D. G. Luenberger, "Time-Invariant Descriptor Systems," *Automatica*, **14**, 473–480, 1978.

[99] Milic M. M. and V. B. Bajic, "Stability Analysis of Singular Systems," *Circ. Sys. Signal Proc.*, **8**, 267–287, 1989.

[100] R. Marino and P. V. Kokotovic, "A Geometric Approach to Nonlinear Singularly Perturbed Control Systems," *Automatica*, **24**, 31–41, 1988.

[101] N. H. McClamroch, "Singular Systems of Differential Equations as Dynamic Models for Constrained Robot Systems," in *Proc. of the IEEE International Conf. on Robotics and Automation*, San Francisco, CA, 21–28, 1986.

[102] N. H. McClamroch, "Feedback Stabilization of Control Systems Described by A Class of Nonlinear Differential-Algebraic Equations," *Sys. & Contr. Lett.*, **15**, 53–60, 1990.

[103] B. G. Mertzios, M. A. Christodoulou, B. L. Syrmos, and F. L. Lewis, "Direct Controllability and Observability Time Domain Conditions of Singular Systems," *IEEE Trans. Automat. Contr.*, **33**, 788–791, 1988.

[104] P. Misra and R. V. Patel, "Computation of Minimal-Order Realizations of Generalized State-Space Systems," *Circ. Sys. Signal Proc.*, **8**, 49–70, 1989.

[105] C. H. Moog and G. Glumineau, "Le problème du réjet de perturbations measurables dans les systèmes non-linéaires-applications à l'amarage en un seul point des grands pétroliers," in *Outils et Modèles Mathématiques pour l'Automatique, l'Analyse de Systèmes et le Traitement du Signal*, I. D. Landau, editor, volume III, 689–698, 1983.

[106] Peter C. Müller, "Linear Mechanical Descriptor Systems: Identification, Analysis and Design," in *Preprints of IFAC Conf. on Control of Industrial Systems*, Belfort, France, 501–506, 1997.

[107] R. W. Newcomb, "The Semistate Description of Nonlinear Time-Variable Circuits," *IEEE Trans. Circ. & Syst.*, **28**, 62–71, 1981.

[108] H. Nijmeijer and A. J. van der Schaft, *Nonlinear Dynamical Control Systems*. Springer-Verlag, New York, 1990.

[109] R. E. O'Malley, Jr., *Singular Perturbation Methods for Ordinary Differential Equations*. Springer-Verlag, New York, 1991.

160

[110] C. C. Pantelides, "The Consistent Initialization Of Differential-Algebraic Systems," *SIAM J. Sci. Stat. Comput.*, **9**, 213–231, 1988.

[111] C. C. Pantelides, D. Gritsis, K. R. Morison, and R. W. H. Sargent, "The Mathematical Modelling of Transient Systems Using Differential-Algebraic Equations," *Comput. Chem. Engng.*, **12**, 449–454, 1988.

[112] P. N. Paraskevopoulos and F. N. Koumboulis, "Observers for Singular Systems," *IEEE Trans. Automat. Contr.*, **37**, 1211–1215, 1992.

[113] G. M. Peponides and P. V. Kokotovic, "Weak Connections, Time Scales, and Aggregation of Nonlinear Systems," *IEEE Trans. Automat. Contr.*, **28**, 729–735, 1983.

[114] G. M. Peponides, P. V. Kokotovic, and J. H. Chow, "Singular Perturbations and Time Scales in Nonlinear Models of Power Systems," *IEEE Trans. Circ. & Syst.*, **29**, 758–767, 1982.

[115] L. R. Petzold, "Differential/Algebraic Equations Are Not ODE's," *SIAM J. Sci. Stat. Comput.*, **3**, 367–384, 1982.

[116] J. W. Ponton and P. J. Gawthrop, "Systematic Construction of Dynamic Models for Phase Equilibrium Processes," *Comput. Chem. Engng.*, **15**, 803–808, 1991.

[117] A. C. Pugh, G. E. Hayton, and P. Fretwell, "Transformations of Matrix Pencils and Implications in Linear Systems Theory," *Int. J. Contr.*, **45**, 529–548, 1987.

[118] R. E. O'Malley, Jr., "On Singular Singularly-Perturbed Initial Value Problems," *Applicable Analysis*, **8**, 71–81, 1978.

[119] R. E. O'Malley, Jr. and J. E. Flaherty, "Singular Singular Perturbation Problems," in *Singular Perturbations and Boundary Layer Theory*, C. M. Brauner, B. Gay, and J. Mathieu, editors, number 594 in Lecture Notes in Mathematics, 422–436, Springer-Verlag, Berlin, Heidelberg, 1977.

[120] P. J. Rabier and W. C. Rheinboldt, "A General Existence and Uniqueness Theory for Implicit Differential-Algebraic Equations," *Differential & Integral Equations*, **4**, 563–582, 1991.

[121] P. J. Rabier and W. C. Rheinboldt, "A Geometric Treatment of Implicit Differential-Algebraic Equations," *J. Diff. Equations*, **109**, 110–146, 1994.

[122] S. Reich, "On a Geometrical Interpretation of Differential-Algebraic Equations," *Circ. Sys. Signal Proc.*, **9**, 367–382, 1990.

[123] S. Reich, "On an Existence and Uniqueness Theory for Nonlinear Differential-Algebraic Equations," *Circ. Sys. Signal Proc.*, **10**, 343–359, 1991.

[124] J. G. Renfro, A. M. Morshedi, and O. A. Asbjornsen, "Simultaneous Optimization And Solution Of Systems Described By Differential/Algebraic Equations," *Comput. Chem. Engng.*, **11**, 503–517, 1987.

[125] W. C. Rheinboldt, "Differential-Algebraic Systems as Differential Equations on Manifolds," *Math. Comput.*, **43**, 473–482, 1984.

[126] W. C. Rheinboldt, "On the Existence and Uniqueness of Solutions of Nonlinear Semi-Implicit Differential-Algebraic Equations," *Nonlinear Analysis, Theory, Methods & Applications*, **16**, 647–661, 1991.

[127] A. Saberi and H. Khalil, "Stabilization and Regulation of Nonlinear Singularly Perturbed Systems–Composite Control," *IEEE Trans. Automat. Contr.*, **30**, 739–747, 1985.

[128] J. M. Schumacher, "Transformations of Linear Systems under External Equivalence," *Linear Alg. Appl.*, **102**, 1–34, 1988.

[129] D. N. Shields, "Observers for Descriptor Systems," *Int. J. Contr.*, **55**, 249–256, 1992.

[130] L. M. Silverman, "Inversion of Multivariable Linear Systems," *IEEE Trans. Automat. Contr.*, **14**, 270–276, 1969.

[131] A. N. Tikhonov, "On the Dependence of the Solutions of Differential Equations on a Small Parameter," *Mat. Sb.*, **22**, 193–204, 1948.

[132] J. Unger, A. Kröner, and W. Marquardt, "Structural Analysis of Differential-Algebraic Equation Systems – Theory and Applications," *Comput. Chem. Engng.*, **19**, 867–882, 1995.

[133] G. C. Verghese, *Infinite-Frequency Behavior in Generalized Dynamical Systems*, Ph.D. thesis, Dept. of Electrical Engg., Stanford University, Stanford, CA, 1978.

[134] G. C. Verghese, B. C. Levy, and T. Kailath, "A Generalized State-Space for Singular Systems," *IEEE Trans. Automat. Contr.*, **26**, 811–831, 1981.

[135] L. van der Wegen and H. Nijmeijer, "The Local Disturbance Decoupling Problem with Stability for Nonlinear Systems," *Sys. & Contr. Lett.*, **12**, 139–149, 1989.

[136] L. Xiaoping and S. Čelikovsky, "Feedback control of affine nonlinear singular control systems," *Int. J. Contr.*, **68**, 753–774, 1997.

[137] W. Yim and S. N. Singh, "Feedback Linearization of Differential-Algebraic Systems and Force and Position Control of Manipulators," *Dynamics and Control*, **3**, 323–352, 1993.

[138] E. L. Yip and R. F. Sincovec, "Solvability, Controllability and Observability of Continuous Descriptor Systems," *IEEE Trans. Automat. Contr.*, **26**, 702–706, 1981.

[139] Z. Zhou, M. A. Shaymann, and T. J. Tarn, "Singular Systems: A New Approach in the Time Domain," *IEEE Trans. Automat. Contr.*, **32**, 42–50, 1987.

Index

Feedback control, of nonregular DAE
systems, 52–72
algorithm for derivation of equivalent DAE
system, 53–58
dynamic state feedback regularization,
58–64
notation, 71–72
Greek letters, 71
math symbols, 72
Roman letters, 71
preliminaries, 52–53
state-space realizations and controller
synthesis, 64–70
Feedback control, of regular DAE systems,
31–51
derivation of state-space realizations, 32–45
algorithm for reconstruction of
algebraic variables, 34–43
state-space realizations of DAE system,
43–45
notation, 50–51
Greek letters, 50
math symbols, 51
Roman letters, 50
preliminaries, 31–32
state feedback controller synthesis, 46–49
controller synthesis, 47–49
preliminaries, 46
problem formulation, 46–47
Feedforward
compensation, 149, 150
/dynamic state feedback compensator, 147
/feedback
compensation, 76
controller, 148, 152
law, 83
/state feedback regularization, of
nonregular systems, 81
Feed reactant temperature, 110
Flow rates, in network, 105
Fluid(s)
flow
Newtonian, 29
rate, 109
incompressible, 29
Frobenius theorem, 99, 127

G

Gas

behavior, ideal, 28
phase reactor, 28

H

Heat
output, 123
of reaction, 110, 123
transfer
coefficient, 15
rate, 15
Heating fluid temperature, 110

I

Impulse observability, 4
Index-three model, 151
Inlet vapor stream temperature, 123
Input(s)
/output behavior, 68, 79
smooth, 59
transformation, 58
vector, 85
manipulated, 46, 117
new, 58

J

Jacket
temperature, 110, 111, 112
volume, 110
Jacobians, 8, 94
Jordan block, 3

L

Latent heat of vaporization, 123, 141
Lie bracket, 119
Lie derivatives, 33, 35, 54
Linear systems, 2
Liquid
density, 110
molar density, 141
phase molar density, 123
Lyapunov techniques, 2

M

Mass transfer
coefficients, 142
correlations, 20